Thought Experiments – An Inexpensive Tour of the Entire Universe

ISBN: 9781945811067

Contact: s.vedantham at gmail dot com

Cover Design: R. Subburayan

First Edition: July 2017 (Tamil)

Second Edition: April 2020 (English - Paperback)

Third Edition: August 2022 (English - Hardcover)

சொல்வனம்

Solvanam Publications

THOUGHT EXPERIMENT

An Inexpensive Tour of the Entire Universe

SUNDAR VEDANTHAM

சொல்வனம்

To Maya

Table of Contents

Preface

Whenever we describe a new scenario to run a thought experiment, we want readers to delve into it, immersing themselves fully in the scenario being described. But it is natural for the mind to wander in different directions to understand what is being proposed. To grasp and enjoy each thought experiment, it is important to slow down that thought process to mindfully understand the notions being presented or debated. The mode of writing is intentionally precise and focused to aid that process. So, instead of reading this book like a novel or a collection of short stories, please see if you can read each chapter slowly and discuss the experiments with family & friends. It is my sincere opinion that such an approach will enrich your reading experience. Asking whether this mode/model is suitable to this twitter world that is forcing us to express everything in 140 characters is quite appropriate. I still would request you to try, in the hope that it will turn out to be worthy of your time.

All the profits earned from the sale of this book will be donated to the Solvanam.com website – a non-profit venture run by volunteers trying to raise the level of discourse in Tamil.

Sundar Vedantham

--------- o0o ----------

The Notion of Thought Experiments

In law a man is guilty when he violates the rights of others. In ethics he is guilty if he only thinks of doing so.

— *Immanuel Kant*

There are very complex and expensive experiments such as Large Hadron Collider, International Space Station (ISS) and International Thermonuclear Experimental Reactor (ITER) that are being put together gobbling up billions of dollars in construction and operation costs. They are balanced out on the other end by very interesting thought experiments that cost nothing but your imagination and time. One can certainly say both kinds of experiments are attempting to answer equally complex questions that surround our lives, while serving very different purposes. But one can't help admiring the elegance and simplicity of mere thoughts that provide extraordinary answers and insights into vexing questions. The idea of analyzing issues by posing questions has been around for a while. That practice has led to a slew of thought experiments (TEs) that span the gamut of practically every field in the world. This book is an attempt to look at TEs that span Science, Mathematics,

Psychology, Economics, Politics, Medicine, Religion, Ethics, Law, Philosophy and other such domains to gain a panoramic view of the issues they place in front of us for our own analysis and understanding. Certainly, you would have heard of some of these thought experiments before. As long as you haven't heard every one of them before, be warned that your sleep patterns may get spoiled for a while! Let's start with one on ethics.

Trolley Experiment

Imagine a train track on which a trolley is rolling along. Unfortunately, there is no driver controlling that trolley. There are no passengers either. Somehow it becomes unmoored from wherever it was supposed to be and has started rolling on the track. You are near the track and notice this hazard. Panic stricken, you look down the train track and realize that there are five deaf workmen working on the track unaware of this trolley rolling down towards them. Unless you jump into action and somehow stop the trolley, it is bound to kill them all in the next minute or two!

Luckily there is a lever close to you that you can use to switch the trolley's track over to the other side. With a sigh, when you reach for the lever, you notice just one deaf worker working on the alternative track. So, if you switch the trolley to the second track, you are going to end up killing that one worker. You need to make the decision within the next few seconds to pull the lever or leave it alone. Will you pull the lever or let it be? You are not allowed to dodge the core question by giving answers like I will run ahead of the train and save them all. The goal here is to evaluate available choices to determine which one appears more acceptable.

Investigators have described this scenario to thousands of people all over the world and asked the same question. Most of those who answered, in fact more than 80% of them, answered saying they will pull the lever since saving five people in place of one is better any day.

Philippa Foot

A British philosopher named Philippa Foot created this TE circa 1967. It is easy to make small variations in this story and ask the question again to understand how people think. As an example, instead of being next to the lever that can switch the trolley's track, imagine you are on top of a bridge watching this trolley coming down. There is another fat man standing on the same bridge close to the edge. The only option you have, to save the five people working on the track, is for you to push this man onto the track in front of the trolley. When the trolley hits the man, it will kill him and then stop, thus saving the five people down the track. Will you push the man on to the track?

When the question is changed this way, most participants trying to evaluate the choices become hesitant. It doesn't matter whether the person answering the question is a man or a woman, young or old, rich or poor, theist or atheist. Irrespective of which nation, religion, tribe they belong to, invariably everyone says that they will not push the man over even if it results in the death of five people on the track. If you look at the two scenarios logically, in both cases we are killing one to save five. If so, why do we feel comfortable making the choice of pulling the lever in the first case but feel uncomfortable about pushing the man in the second case? Participants who took part in the experiment are usually not able to articulate a clear reason to justify their change of heart. But they don't seem to struggle with the decision and are quite clear as to what they will or will not do. Ethicists explaining the difference point out that in the first scenario either five or one is bound to die. But in the second story, the fat man watching the trolley from the top of the bridge initially has no connection to the accident that is about to happen. The connection comes into the picture only when you push him from the bridge onto the track. So, even though we may be saving five people in this scenario, sacrificing someone who is not at all connected to the accident to achieve that end doesn't ring right in most people's thought process.

This is not just an academic experiment totally unrelated to reality. Such choices do show up in real lives very often. At the end of the Second World War, when the US dropped two atom bombs on Japan, the reason given was that to control the madness of Japan, and to end the world war saving several thousand lives, it was necessary to destroy one or two cities like Hiroshima and Nagasaki, even though they may have held a lot of innocent people. We are not debating the veracity of that argument but just pointing out how such huge world events are connected to this TE. Those atomic bomb drops were equated to pulling the lever to switch the trolley's track in the first scenario. Dropping the same bombs in some other cities, say in Africa, killing people there to scare Japan is equated to the second scenario of pushing the man from the bridge. Though the end result might have been the same, this experiment emphasizes the importance of considering how you get there.

In the age of self-driving cars, when a car is about to hit a pedestrian, what the car should do can be precisely coded into the software. If the car is about to hit five pedestrians illegally crossing the road, should the software force the car to swerve and hit one person legally walking on the side of the road, so that four lives can be saved in the end? We can simplify the scenario even further. When there is just one pedestrian illegally crossing the road, should the software hit the brake hard so as not to hit that pedestrian, even if it will take out the teeth of the car's passenger? If it does, since the pedestrian is saved, that good deed may not be noted. But the passenger who lost his teeth may sue the car manufacturing company for damages. While such decisions are currently being made by the driver and the speed of their reflexes, self-driving cars can implement the priorities and decisions much more consistently, warranting pre-agreed and approved priorities. If you would like to contribute your thoughts to this debate, you can visit http://moralmachine.mit.edu/, a website that is surveying and curating the opinions of the people all over the world.

We can even think of scenarios pertinent to organ transplant that can be mapped onto this thought experiment. Let us say there are five patients near death, awaiting heart, lung, kidney and couple of other transplants. Can we kill a healthy human being to harvest his five organs so that we can save those five patients awaiting organ transplants? "No Way, Jose!" will be the obvious answer here. That is why we can't accept the notion that the end justifies the means under all conditions. But naturally, if a human being is going to die or is brain dead, harvesting his/her organs to save multiple patients awaiting transplants is unquestionably a major achievement of modern medicine that we are all very proud of.

We can create so many variations of this trolley story. In one scenario, instead of placing five people on the track, we can place just two, or even one person. Perhaps we can swap male workers with one or more women and children. When we make such changes, answers do change. But the changes appear to be strikingly similar across demographic, regional, linguistic, and religious boundaries! Philosophers have pointed out this similarity to claim that human beings don't need religion or the fear of God punishing us in an afterlife to teach us morality, since notions like egalitarianism, ethics and respect/love for fellow human beings are all innate across societies and civilizations. What do you think? You can try to discuss these scenarios with your family and friends and tabulate the results.

Heinz's Dilemma

Heinz is a blue-collar worker, earning $50,000 a year. His wife is diagnosed with an unusual form of cancer. Heinz can see with his own eyes how she is suffering. Doctors have given her just two more months to live, unless someone can find her a miracle cure. Searching through every avenue accessible to him, Heinz comes to know that there is a new drug that can cure Mrs. Heinz. But the cost of treatment is one million dollars. What do you think Heinz should do?

1. Heinz should just make sure she is as comfortable as she can be until she passes away. The needed drug being so expensive, while Heinz's income being so low is his fate (or God's plan). So, Mrs. Heinz may end up dying, and no one can do anything about it.

2. Heinz should petition/beg the drug company to reduce the price of the drug or donate it to him.

3. Heinz should borrow money from the bank and get the drug to treat her.

4. Heinz should just break into the drug company and steal the medicine. Since his wife's life is at stake, he has every right to get hold of the drug. If he gets a prison sentence later for that crime, he should accept it.

5. Heinz should steal the medicine. He has the moral authority to demand that drug, so he should resist imprisonment and should escape the police.

6. All the listed choices are incorrect or not fair. We need to find a new alternative.

Lawrence Kohlberg

An American ethicist named Lawrence Kohlberg posed this question. Psychologists say that as your answers evolve from 1 to 5, they indicate how you are breaking free of conventional thinking. Citizens of a society refusing to heed the law or taking the law in their own hands will certainly affect the society adversely. But Kohlberg argues that as we grow from early childhood into full adulthood and then old age, we slowly evolve step-by-step. Initially we are taught to, and do respect all the rules, satisfying the expectations of elders, thus pleasing them. Then we try to live in harmony with the community surrounding us. Subsequently, our innate need to respect human dignity starts to push us into opposing laws and rules we don't consider fair. Kohlberg thinks these are normal evolutionary steps in most human beings as they grow old.

What if we change the story a bit and state that Mrs. Heinz's issue is not fatal cancer, but just a recurring headache she suffers from? We can go even further. Let us say Mrs. Heinz is healthy, and so there is no disease to cure. But this drug will just elevate her mood and keep her feeling happy consistently. You may bristle at such a question with, "What a nonsensical question is this? Is there no difference between a lifesaving drug and a recreational mood elevator?" For example, most of us will be willing to condone a poor homeless person stealing a loaf of bread out of hunger. But it is hard to condone a rich guy who has four cars, for stealing the fifth one, even if the fifth car was stolen from a billionaire who owns 100 cars. Thus, we have a sense of difference between what we consider as want vs. need. There was a recent example of this phenomenon. The debate was whether American health-insurance policies should cover contraceptive pills for women. Many insurance companies and organizations argued that they need not be covered. But many women's organizations argued that they should be covered, as it is often a necessity for women to maintain good health. They also pointed out that the same healthcare organizations routinely cover Viagra prescriptions under the health insurance policies they have been selling for years, and wondered why there is a discrimination between what is covered for men vs. women. Though this may sound like a liberal vs. conservative debate, we can see a clear connection between these questions and Heinz's dilemma.

Even those that own good health insurance policies often struggle with issues at the other end of the spectrum. After many years of life, a patient's body may be failing in the hospital, but thanks to the latest machines and technologies that are available now, it is possible to keep the patient alive for months or years at enormous costs. Since health insurance can pay for these equipment and procedures, whether to keep someone artificially alive though the quality of life will be next to nothing becomes a serious debate. Wouldn't the patient and close relatives as well as the system be better off if we ensure that patients on their deathbeds do not unduly suffer pain or other quality of life issues by allowing them to pass away in peace? This is a recurring question on a daily basis in every ICU in every hospital in the US.

Irrespective of whether we care about the cost incurred by the society or the insurance company when one's life is artificially prolonged, we should be able to appreciate the emotional costs our nearest and dearest will be forced to bear under such circumstances. So, doing a simple thought experiment might help. Imagine yourself being under various medical conditions that preclude you from making your own medical decisions. Would you like your caregivers to do everything possible to keep you alive or would you like them to let you go? What if the probability of whether you will recover enough to lead a normal life is high/low? What if you will recover but will lose the quality of life (e.g. you will be in constant pain or will need 24x7 help to breathe/bath/eat/ease yourself)? Collecting your thoughts on paper is effectively the process of creating a Living Will. Writing one's Last Will and Living Will is a task we all should take up so that our relatives do not have to face the dilemma of choosing between keeping us alive while in pain or in a coma or letting us die peacefully.

--------- o0o ----------

Everyday Encounters

To make no mistakes is not in the power of man; but from their errors and mistakes the wise and good learn wisdom for the future.

— *Plutarch*

The Trolley Problem and the Heinz Dilemma we encountered in the previous chapter assess the strength of our moral compass. Even if we skip such philosophical assessments and come down to a lower level, we start to realize that TEs are intertwined in everyday life as well.

Ship of Theseus

Plutarch

In the recent past there was a movie named *Ship of Theseus* that explored this notion. About two thousand years back, Greek writer Plutarch posed a question. There is this shipping tycoon named Theseus, who has a well-built brand-new ship. As the ship is key to his thriving import/export business, he took good care of the ship with annual maintenance. Over the years as the wooden ship's planks started wearing out, he kept replacing the failed planks. In a decade or two, in addition to individual planks, if all the components including the ropes, the mast, and sails have all been replaced, does Theseus currently own his original ship or a different one? This is the question Plutarch posed. If none of the original components remain in his new vessel, isn't it indeed a new ship? If we concede that it is indeed a new ship, when did it become a new ship? Perhaps when 51% of the original ship was replaced? Or is it only when the last piece of the old ship is replaced, thus completing the new ship's parts being 100% different? Let's argue that changing parts do not matter and it is still Theseus' old ship. Now let's say the dump yard owner who received the discarded parts of Theseus' old ship secretly rebuilt the ship piece by piece as he received those discarded parts over the decades. Obviously, this reconstructed old ship with all its rotten innards will certainly not be seaworthy. But now, out of these two ships that exist in the world at the same time, which one can be called Theseus' ship?

Even after 2000 years, it is hard to find an easy answer to this question, though it seems to pop up often in everyday life. Let us say four musicians get together to form a band like The Beatles. The band becomes very famous and successful, and they earn a lot of money. As time passes, as the original members of the band age out and retire, new musicians join the band and fill in the holes. Let us say after a while all four original team members have been replaced by new musicians. Now should the band be recognized as the same old band or is this a different band? If a new contract gets them a lot of royalty as their songs sell well in MP3 format, how much royalty should flow to the original musicians? Questions/issues like these get argued out in our courts every year.

Let us leave music groups alone and look at individuals living their own personal lives. We know of heart and kidney transplants that have been going on for eons. Replacing faulty knees and hips with artificial ones made of metal alloys is fairly common now. If aging hearts don't maintain the rhythm perfectly, pacemakers are available now to help maintain the pace. In 2005, physicians in France demonstrated a partial face transplant. By 2010 full-face transplant became possible. So, now full-face transplants have been completed successfully in multiple countries. It is only natural that we will have access to more and more artificial body parts in the future that will perform better than the originals. When we continue to replace the parts of a human being one-by-one, eventually, does that person lose his or her identity? Perhaps for now, we can answer this question by saying that as long as you have your brain and memories intact, you are indeed you. What if brain transplant becomes possible in the future? We don't even need to go that far. If personal memories and the consciousness of yourself is the real you, what about a patient who is losing his memory to Alzheimer's? As that cruel disease robs him of his memories of his life, should we declare that he is no longer the same person? If so, at what point of memory loss? If not, why not?

Whether women should have the right to abortion or not is a long running feud in American society. In particular, a large percentage of religious Christians tends to argue that aborting a fetus at any stage of the pregnancy is morally wrong. They also tend to vote for politicians who oppose abortion. This electorate that calls itself Pro-Life usually tends to support easy access to guns and death penalty, though these decisions may also result in loss of lives. This cognitive dissonance is noted by a Pro-Choice American politician who quipped that the Pro-Life groups seem to think that life starts at conception and ends at delivery since they don't seem to care much about life afterwards.

As per the views of the Pro-Life groups, even if a woman has the right to abortion as provided by American legal system, since that process terminates the life of a fetus that is not capable of fighting for its own rights, the sin and responsibility of committing that murder is on the woman who goes through the procedure. So, they argue that even if a pregnancy is a result of rape or incest, the woman carrying the fetus should consider it as God's Will and forego abortion. An American Philosopher

Judith Johnson

named Judith Johnson proposed a thought experiment related to this idea around 1971.

In this experiment, a world-famous violinist is in a coma due to a tricky ailment. By a freak medical miracle, the medical community has found a unique procedure to help address the violinist's ailment with your help. As per this plan, one day when you are asleep, fans and physicians bring that violinist still in coma to your home and attach him to you physically, using a machine. You are shocked to find this setup when you wake up. If you let him be attached to you in this fashion for the next nine months, he will take part of your nourishment to survive and get healthy and will eventually disengage. If you decide not to go through with this setup and snip the connection, he will certainly die.

What will you do? Will you wait around for nine months to support him and help him survive? Or will you decide that saving his life is not your responsibility and cut the cord? Since that musician was in a coma when he was attached to you, he was just like a fetus attached to you. So, if one who hasn't done anything wrong has a right to live, it implies that you don't have the right to terminate your support. If you have the right to terminate your support, it implies that his life can be taken away for no fault of his own. Which side of the argument is right?

What Johnson is trying to argue is this. Though the violinist in a coma has a right to live, and though you are the only one who could keep him alive, whether you decide to give him what he needs or not is completely your prerogative. When you snip the cord, you are simply executing your right. If that results in the death of the violinist, it is not your responsibility ethically or legally. Thus, you have every right to decide what you want to do from your side. Do you agree with this argument?

Thanks to some natural changes in the human body or new developments in science, if men become capable of conceiving and delivering babies, what will happen? Women on the Pro-Choice side joke that the very next day all the male politicians that want to ban abortion now will do a somersault on their position and argue that abortion is a God-given right to human beings that no one should be allowed to take away!

Spider in the Toilet

Thomas Nagel is a contemporary American professor of philosophy. When he went to use the urinal in the Princeton University where he worked, he saw a spider hanging there pathetically. It looked like the spider didn't know how to walk out of the slippery porcelain surface of the urinal. To Nagel it looked like the spider somehow got stuck unwillingly, with a disgusting rain of urine pouring over it often. After using the facility, Nagel returned to his office feeling sorry for that spider. When he returned to the toilet the next day, the spider was still stuck on that same spot continuing to look pathetic. It is not clear as to how it continued to survive there or from where it got its food. Not being able to take it anymore, Nagel found a paper napkin and took it close to the spider, trying to nudge it slowly. It did climb on to the napkin allowing Nagel to get it away from the urinal basin. Nagel then left the spider in a clean dry place nearby, telling it to move on with its life and live happily ever after.

Thomas Nagel

Feeling smug, he returned to his office, but a little later, when he went back to the toilet, he noticed the spider standing pat in the same location. After asking the spider, "What? Are you not ready yet to move on with your life?" he walked away. A day or two later when he went back to the toilet, he found the spider dead in the same place where he had left it. Nagel wrote about this incident later asking the readers to consider it from a thought experiment point of view. However sure or moved we may feel, 'Is interfering in someone else's life the right thing to do, even if we are convinced that it is for their own good?' is the question he poses. Perhaps living in that toilet bowl is the only way of life the spider knew. Nagel wonders whether, instead of helping the spider by giving an improved life to it, he had unintentionally killed it, without understanding how it fed itself.

If saving the life of this spider which was 'obviously living a miserable life' could go wrong, how can we be sure that 'saving' someone by nudging them to go through religious conversion, or 'civilizing' primitive tribes are all indeed good deeds? When questions are raised about domesticating pets and raising small birds like parakeets in cages, those who support that practice argue that these pets are born/hatched to be raised this way and letting them go free will actually end up killing them as they are sure to become lunch for a cat outside. Opponents argue that breeding such pets that cannot take care of themselves is inherently wrong, since it is equivalent to forcibly interfering with their lives to begin with. Sorting out these questions in a logically consistent manner without falling victim to any emotional bias is innately difficult.

Two other thought experiments follow this trail. The first one asks us to imagine a world that has given up eating meat and has completely converted to vegetarianism. Since there will be no need for all the animals that are currently being raised to be eaten by human beings in that world, they will no longer be raised. As a result, dozens of species like chickens or cows may even go extinct. So, in a way one can argue that, more than human beings, the animals that may go extinct have a bigger interest in human beings continuing to be meat eaters, so that they can preserve themselves. Opponents call this argument bunk. Though various species may have a Darwinian interest in preserving their lineage, if we buy this argument, then despite knowing fully well that there is no good future, giving birth to millions of animals and forcing them to live miserable lives should be a good thing. If we agree with that take, then it should also follow, that raising human beings to be thrown off bridges to stop unruly trolleys that are rolling down the track is acceptable as well.

John Rawls

There is yet another view that opines that we cannot treat animals and human beings the same way. That leads us to the next TE introduced by a philosopher named John Rawls. In his book named *Theory of Justice*, he tackles the process of creating a political system and a set of laws that could work well for any type of society at any period of time. To be able to design such a society, Rawls suggests a TE called *Original Position* where he asks us to wear a veil of ignorance that will allow us to forget our current social status, standard of living, the country we live in, laws and rules we are used to, thus setting us up in a common baseline. Once you reach this mental state, you need to choose a societal structure and a set of laws that you think will fit a society to be created anew, that you would like to live in. Once your choices are made, what kind of a person you will be and where you will live in that society will be revealed to you.

Let us say you support a societal structure that provides capital punishment for drug peddlers. You may know that as of now many countries in the world have such laws in their books.

After you make that choice, when your avatar's characters are revealed to you, how will you feel if you turn out to be a person with some mental challenges who is forced to sell drugs on the street by a ruthless Don you are quite afraid of?

How we think, what rules & laws we support are all based on our current background, and our current society's values is the point being made here. That is why rules that appear correct to the Queen of England, a slum dweller in India and a tribal woman in Africa may not have much in common. So, if we are forced to pre-choose a set of laws and a social structure before we get to know our position in the society, it is plausible that such choices will be conducive for any society to thrive in any period. When our character is revealed to us after the choices are made, it might not even turn out to be that of a human being but of an animal being raised to eaten (like a chicken) or perhaps a chimpanzee that resembles human beings in so many ways. How will we feel then? Imagine how our choices will change once we understand the possible range of choices of such avatars in the end. The main kernel of the storyline in the movie series, 'Planet of the Apes' is indeed this thought experiment that places the shoe on the other foot, so that we will understand the way the game is played much better. If you haven't seen those movies, it might be worth checking out at least the first one in the series.

--------- o0o ----------

Note: When this section was published on the Solvanam.com site, there were two reader comments that appeared very relevant and interesting to the TEs discussed on this chapter. They are reproduced here verbatim:

Reader Mr. Krishnan said:

Like the story of "spider in the urinal", there is a much more interesting story in Mahabharatha. Mukkur has told this is in his Kurai Ondrum Illai. Once a shishya (student) did some major mistake and the guru punished him to be born as a pig. Before that transformation happened, the shishya realized his mistake and asked his friends to chase the "pig" (i.e. when he would be born as a pig) and kill him so he could be born back as a human being as the guru had promised. Friends agreed and after the transformation, they got to know from the guru where their friend-pig was to be born. After going the information, they searched and found out their friend-piglet; the friend-piglet realized the situation but when the friends tried to kill him, he started running for life, shrieking that he loved being a pig, he loved his mom and dad so much, and he would not give up that life for anything, etc. Compared to the "spider" incident, where the author only vicariously felt the "feelings" of spider, here the same person was set to be born as a pig to realize the various perceptions of various lives and livings. Mahabharatha has hundreds of stories carrying heavy loads of deep thoughts. Another example of such stories is what Sarvapalli Radhakrishnan reportedly told the greatest scientist, Einstein, how a king (in Mahabharata) went to heavens, spend just a morning and came back to his kingdom to see three generations had passed on, relating to the relativity theory of Einstein.

Reader Mr. Baktha said:

Ship of Theseus: More recent reference to this concept is the fact that periodically cells in our body die and new cells are created and over a period a few months/years our body has a set of cells, none of which existed few months/years back. So, what is this body that we are talking about? This concept is also increasingly being used to give the much needed mental strength to people suffering from chronic diseases like say cancer. To tell them that it need not necessarily be terminal and that there is a possibility for cure. The power of the thought, power of hope, sometimes power of ignorance (being bliss) is capable of curing people of even ailments like cancer. Some years back, my sister's m-in-law had some cancer in her stomach. She was not literate. The family decided not to tell her anything about her ailment, other than the fact that she had some kai (boil) in her stomach. She underwent radiation, chemo etc. She lived through next few years. She did go thru physical pain, but she lived happily and eventually died.

Original position – Drug peddler – paradigm shift: Many years back, we had a student who was an orphan. I came to know that he is an orphan when I had an inquiry from the State social welfare department about him. When I started doing more background checks on him, I came to know that he used to sell Ganja (Marijuana) to fellow students. He also carried a knife with him almost always. I was angry. But only on further probing, did I learn the following: His father killed his mother in a drunken brawl. His "mama", that is mother's brother, killed his father in an act of revenge. Result: This fellow and his brother turned orphans. He was selling drugs and doing such things only for his very survival! He came to final year and was not getting placed in the first few rounds. We kept communicating with him, giving him confidence. Then, one day, it happened. He got a job in a good core company. He came to my room to thank but soon turned very emotional. When he gathered himself, I asked him, "What are your future plans?" to which he replied, "My younger brother has to study." Subsequently, he joined the company, was deputed to Japan and was doing extremely well.

Experimenting with Computer Science

Computer science inverts the normal. In normal science, you're given a world, and your job is to find out the rules. In computer science, you give the computer the rules, and it creates the world.

— *Alan Kay*

Let's move away from philosophical or questions related to daily life and dive into thought experiments in computer science. Our interest will be particularly focused on three experiments related to artificial intelligence within computer science.

Turing Test

The twentieth century British mathematician named Alan Turing left his impressions all over the field of computer science though he passed away at the young age of 41. He is considered the father of theoretical computer science by many in the field. In 1950, Turing published a paper on machine intelligence. The purpose of the paper was to discuss computers' ability to think like human beings. The elegant thought experiment he proposed to address the issue was subsequently named the Turing Test and is being actively discussed even today. In this experiment there is a human being at a computer terminal in one room and either a computer or another human being in a second room. Suppose a computer is in the second room. If the human being in the first room interacting through a terminal cannot determine if the entity on the other end is a human being or a computer, then Turing's suggestion is that we can consider that computers can think and function like human beings. In other words, when computers interact with us, if they can mimic all the abilities as well as the shortcomings of human beings to such an extent that they are indistinguishable from real human beings, then Turing asserts that the level of AI has grown to a level equivalent to

Alan Turing

that of human beings.

After examining the rate at which AI development was progressing in the 1950s, Turing assessed that by the turn of the 20th century, computers would be able to convince more than 30% of the population interacting with them that they are indeed human beings. This should be interpreted as, 'he expected the sophistication of the emulation to grow significantly rather than the intelligence of the general population to decline.'

Unlike the thought experiment we discussed in the first chapter, this thought experiment is bloodless. So, the software engineering community has been trying consistently for decades to see if this state can be attained in reality. As per a recent announcement, 30% of the judges that interacted with a remote computer came away convinced that they were interacting with a teenager from Ukraine. Based on the touchstone provided by Turing seventy years back, one can argue that we have reached the point where computers have managed to pass the Turing Test, thus marking a milestone. But an honest inspection will reveal that till now we haven't created computers that really think. What we have managed to do is reach this milestone depending on extensive jugglery, obfuscation, smoke and mirrors techniques. That's the hard truth. But it is also true that both the engineers and computers have been running in that direction by continuously fine-tuning the approach each day.

You might have heard about the famous US television quiz show called Jeopardy, in which contestants provide their answers in the form of a question. In 2011, IBM offered their computing system named Watson as a Jeopardy contestant to compete against two of the best known human Jeopardy champions. Watson crushed the other two contestants, clearly demonstrating its ability to understand colloquial terms, being able to search through its knowledge base for answers that are very context sensitive and doing it all in seconds to press the buzzer in time, after assessing its confidence level in an answer it is able to generate, that too in the form of a question! The famed champion Ken Jennings who competed and lost to

Watson quipped that he is ready to welcome and obey the computer overlords now. If you haven't seen that two-part Jeopardy show, it will be worth your time to hit the search button on YouTube where it is available for free to watch. It is easy to see that the Jeopardy attempt resulted in a spectacular success, and a more recent Hollywood movie titled *Her* also has the Turing Test notion running in the background as a connecting thread.

While there have been attempts where human beings have pretended to be computers to see if judges can figure it out, there is also a need for a version of Anti-Turing Test in the real world. In this case, computers need to be convinced that they are interacting with real human beings and not computers. Including Solvanam.com, you might have noticed several normal websites posting CAPTCHA puzzles, asking you to solve them when you try to post comments or even login to a site. By asking human beings to answer questions that computers can't easily answer, websites determine if the entity they are interacting with is a real human being and not a robot, using what is called 'Completely Automated Public Turing Test to tell Computers and Humans Apart', i.e., CAPTCHA. While it is annoying for human beings, it helps block automated robots from overwhelming websites.

Alan Turing, being the genius he was, in addition to the notion of the Turing Test, has also delivered a thought experiment setup called Turing Machine to the field of computer science. I still remember my early grad school days, when I came across Turing Machines for the first time, and got it confused with the Turing Test that I had heard of before. It took me a couple of days before I realized that they are two different gifts from one man, to the field of computer science. In that theoretical contraption, there is a tape that is of infinite length and a head that can read/erase/write just 0 or 1. That is about it. With that simple setup, one can analyze the time and space needed to perform every modern computation one can imagine. It planted the seed for a whole subfield in Computer Science named 'complexity theory' and has led to a large amount of understanding of problems and their resource requirements. Thus, if we devise a new number sorting algorithm, analyzing the old and new algorithms using a Turing Machine model will help us understand which algorithm is better.

A web search should provide tons and tons of material on this theoretical machine and the understanding it has led us to. Mathematics, computer science, cryptography, several inventions to beat Nazi Germany during the Second World War, winning marathons are but a few areas where Alan Turing shined within the 41 years of life he had. You may remember the Hollywood movie The Imitation Game from 2014 that showed one sliver of his life related to British efforts to thwart Nazi encryption technology. His life is surprising from so many aspects while simultaneously being sad from many others. If your interests are piqued, a web search is highly recommended.

Do You Know Mandarin?

You are sitting inside a room and you know only English. You have a large book full of rules, enough paper and a pen. A gentleman who knows Mandarin well writes a small short story in Chinese along with a few questions to test the reader's comprehension. He then sends this write-up to you either on a piece of paper or via a computer terminal. Though you don't know Mandarin even a little bit, you accept the input that came in, then look through the large book you have. Following the rules listed out in the book, you meticulously compare each character found in the input letter by letter, look through the book's pages and figure out what the expected response should be. Since the book tells you that if you find this odd bug-like character followed by this funnel cake-like character, you need to draw this leaf & twig-like shape as the response and so on, you are able to copy the shapes as directed, creating a response. It may take a while but the person outside waits patiently until you provide the response. When he reads your response, he attests how perfect your response is. He even thinks some of the philosophical interpretations you have provided are sublime, and so he certifies that you are indeed a Mandarin expert. At this point, can we declare that you know Mandarin?

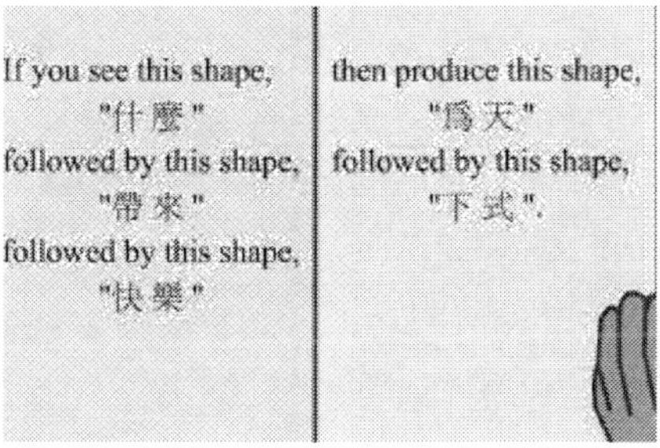

If you think this is a dumb question since you simply followed the rules as laid out in the book and so you are in no way a Mandarin expert, you belong to the Weak Artificial Intelligence or Weak AI club. Scientists in this club say that computers can only emulate such rule following and can never intrinsically comprehend what they are saying themselves as human beings routinely do. There are subtle but important nuances to this discussion. Though you, who is the person in that room, currently do not understand Mandarin, you can learn it in the future in the conventional way people learn a new language. Then you will understand the answers you send out (whether you generated them by following the rule book or wrote them yourself) just like a native Mandarin speaker. Along the same line, the computer as we currently understand only follows rules. But in future there could be more advanced computers that can potentially comprehend what they do and may become self-aware. But the Weak AI campers feel that computers can never ever cross that chasm even in the distant future since that is something innately biological.

Alternatively, people in the Strong AI club argue that computers can certainly become self-aware and understand what they are doing. But even here the discussion is nuanced. They argue that irrespective of how you answer the questions, if you can function as well as a native Chinese speaker, you should be considered as one who knows Mandarin. That team continues to argue that if computers can really function at that level, you need to treat them as sentient beings that have consciousness. In their view, if you break down the mechanics of even human beings, the brain boils down to individual cells and small electrical signals traveling back and forth. Beyond that, the notion of soul and God's image are all just products of sheer nonsense. This line of thinking circumvents the need for fundamentally describing what is meant by terms like 'understanding', 'intelligent', and 'consciousness', and declares that near perfect simulation is equivalent to the original since the original itself can be broken down to the nuts & bolts of the mechanics in the end.

John Searle who suggested this thought experiment belongs to the Weak AI club. Scientists who belong to Weak AI club say that by clever programming, we can make computers do a lot of things that may make them behave like human beings. Human beings may even perceive them as intelligent or even self-aware. That functional state indicates that computers can serve as a good model for human brains. That's it. In other words, these rule following computers have nothing equivalent to the consciousness of themselves or their actions. So, it is the unassailable view of Weak AI scientists that computers can never gain consciousness. Using clever programming techniques computer scientists may even pass the Turing Test. But similar to the human being in that room who doesn't really know Mandarin despite successfully functioning as one who knows, computers never really "understand" anything, but only execute a series of given instructions.

Though Searle originally proposed this Chinese Room experiment as a way to conceptualize and compare/measure the level of AI in various computer programs, it caught the imagination of experts in multiple fields, such as linguists, philosophers, theologians, biologists and medical professionals,

John Searle

who have all jumped into the realm debating the idea vigorously.

Those who oppose Searle present three types of arguments:

1. It is conceivable that the human being in that room doesn't know Mandarin. But we shouldn't be looking at the human being alone while evaluating the question at hand. Besides the person, we need to look at the book that has all the rules, how the questions come in, and how the answers go out. All these elements combined should be viewed as a complete system that "knows" Mandarin. Thus, with the book included, the total system does understand the language in full. So, claiming that there is no real understanding occurring in this system is incorrect.

2. Yes, the system described in the thought experiment does not understand Mandarin. But this is a system intentionally created to be an incomplete one. If you change this system a bit, for example, by adding a camera that serves as eyes of a robot who has enough computing capacity to memorize and execute the rule book properly, when it responds to incoming questions perfectly, we will be forced to agree that the robot understands Mandarin.

3. The third argument doesn't concede any space. Experts that belong to this class argue that what is occurring in that room is indeed understanding of the Mandarin language. If we say it is not, it implies that we haven't

> understood what understanding truly means. Once we get past our confusion as to what understanding is, we will then agree that the man or the system understands Mandarin. Thus, the system understands Mandarin, and it is those who disagree are the ones that don't have the full intelligence yet to answer this question.

Notice how far a computer science thought experiment has progressed to touch a multitude of domains & experts.

Deep Learning

Early approaches to Artificial Intelligence predominantly used Rule-Based Learning. Computer scientists and software engineers code the details of what we expect computers to learn into a set of rules and load the resulting program into the system. Once the program starts running, they let it run on the natural environment where it is expected to function or present it with the data set it is expected to handle and watch how it behaves. Then by closely observing its behavior and comparing it to the expected behavior, they start tweaking the rules as needed to get it to behave as required. This approach worked very well for systems designed to function in well-defined domains, such as a factory floor.

Since the power of the microprocessors (or the CPU) in use and the amount of memory it could access were both very limited, they couldn't advance far beyond this level for quite a while.

Despite these limitations, a scientist named Arthur Samuel who was working for IBM in the 1960s was trying to teach computers how to play Checkers. Initially he taught the system basic rules of what moves are allowed and what moves are not allowed. Right away the system started playing the game flawlessly according to the set of provided rules. But Arthur who was playing against the computer could easily think through few moves ahead and make moves that would lead him to a win. So, he was able to beat the computer consistently since the computer didn't know much beyond what moves are legal/illegal under a given condition.

After spending a few days in this mode, Arthur added another skill to the Checkers program. This skill allowed the computer to assess the probability of success based on each one of the legally allowed next moves. Once this code started working, Arthur let the computer play against itself by running two instances of the same code and went home on a Friday evening. Given this capability, computer was able to learn a lot from each one of its moves, and the resulting win/loss at the end of the game. By the time Arthur returned to office Monday morning, the computer had gone way past his skill level and started beating him consistently. He could never beat the computer program he had written himself ever again.

Two remarkable advances had taken place through this experiment. The first was the new capability computer acquired to gather data by itself, analyze the results and train itself learn to perform better in the next round. (Whether we want to call this process 'training' or 'learning' again leads us to nuanced debates. The term learning may imply something akin to a biological process that happens in the 'mind', while training may be perceived as a mechanical process executed by software.) The second was the program's ability to start with a set of rules human beings provided but master them to such an extent that a human being could no longer compete with the program and win. Thus, this was an early and significant milestone in AI.

Compared to yesteryears, the processing capacity of microprocessors as well as the volume of memory space they can access have increased substantially. So, a new approach of educating computers has come to the forefront in the last decade. In this mode called Deep Learning, the computer is given just a few basic rules initially followed by gobs and gobs of data, letting it make its own inferences based on those inputs and learn what it can.

Since today's computer can process data orders of magnitude faster than a human brain, as this new approach evolves, computers are able to figure out links that exist between different parts of the data and make their own inferences and conclusions, leading to some truly impressive decisions, making them incomprehensible to human brains. This Machine Learning (ML) approach has been unleashed to improve not only self-driving cars and what to sell/show you next based on your past activities, but also improve decision making in medicine/law, predicting future crime spots to prevent crime by deploying police to that location ahead of time, what interest rate you should be made to pay for your credit card debt and so on. We are familiar with the stories of God creating human beings and letting them run wild to see what happens. We can see a parallel in this domain where we have created a new form of intelligence and may be letting it run wild. As the God of this intelligence, the TE here is to determine what our approach towards this intelligence should be. Should we concede that computers will become smarter than us at some point, such progress cannot be stopped by us and so we need to embrace that progress to take full advantage of its capabilities? Do we stand by that decision even if, despite being the creators, we don't fully understand the process? Or should we prevent anything that we can't fully understand within the limits of human capabilities? Different answers to this question can lead us in different directions as the following examples demonstrate.

The reach of this ML technology is nicely illustrated through what happened in Syracuse, NY. It is an old city in the Northeast part of the U.S. So, the network of pipelines delivering drinking water to the citizens is an aging 100+ year old infrastructure. On an average once a day, some part of this waterline breaks somewhere in the city spewing water to the surface, resulting in water being cutoff to that part of the city until the city workers manage to fix it. The city mayor tried asking for state and federal government money to fix these pipelines. She even ran TV advertisements trying to shame the state government hoping that it will get her the needed funds to replace the pipes, but she had no luck.

If you build a large airport or a factory, politicians involved can cut the ribbon, pose for photographs and give interviews to the media proclaiming what a fantastic job they have done. But if you spend time and money fixing broken water pipes running under the city landscape, hardly anyone is going to notice. So, though maintaining the water distribution infrastructure is extremely important to run a city well, no one showed interest in funding this non-glamorous project.

The small-town mayor was clever enough to understand this phenomenon and so decided to attack this problem from a totally different angle. So, she contacted the scientists in the nearby university. After analyzing the problem, that team decided to use Deep Learning to solve this problem. Instead of responding to broken pipe reports and mending them, their approach is to use A.I. to predict which pipe is going to burst before it does. In a way, this approach pushes the program running on the computer itself to get involved in a thought experiment.

As a first step to get this project going, the team gathered all the data that was available about the water supply system in the city hall. There were data to be gathered in decade old dog-eared paper reports stored in cardboard boxes and inspectors' note pads, typewritten memos and more recent data in multiple data bases, CDs and floppy disks. Once all this stuff was collected together, the team designed one database that could be used to collate all the information, making it easy to analyze. Once the database looked consistent and usable, they ran a machine learning algorithm on that data set. While executing the program, they cleverly withheld last five years' worth of data already available to the team, and provided only the first fifteen years' worth of data to the program, asking it to learn the pattern and guess which pipes are expected to burst the next year. When the predictions came in, team compared it to the year 16 reports and determined which predictions are correct and which ones are incorrect. They then fed that information back into the system, such as, 'this one prediction that a main street pipe will burst on 34th Street is correct but the other prediction about the 7th Avenue and Broad Street intersection is incorrect'. With this additional input, the program was rerun using these 16 years' worth of input to predict where the breakages will be in year 17.

Since the team had the year 17 data also in their hands, they could compare the program's predictions with what had really happened and provide another round of corrections to the algorithm. When this process was repeated three more times, the ML algorithm got the hang of it very well and was able to correctly predict where the breakages are going to be in the year 21. So, in that year, before pipe bursts occurred, the repair crew was onsite fixing the potential bursts before they occurred, thus saving the city more than half a million dollars in their first year compared to the previous years. As years progress, the scientists working on this project expect prediction quality to continue to improve. Since the idea can be applied to any city in the world that is facing this infrastructure problem, that team had posted all the code used in the project on GitHub for anyone to download and reuse.

But the same technology has also posed a harder to solve dilemma in front of us. We will reap the benefits of Deep Learning only when we allow the computers and programs to run without any interference. If we tend to guide the learning process or restrict what can or can't be done, we will obviously limit the technique's scope. But if/when we let AI algorithms to run completely on their own, very quickly the decision-making process inside the program becomes opaque to human beings. We are no longer able to understand the "thought process" (of that program) that lead to a given conclusion. So, the question now is whether we are ready as a society to let computers do/decide things that we no longer understand? Are we comfortable believing the system to give us correct decisions and us using them without really understanding why and how? This is a difficult choice to make.

If we are not comfortable living in such a world, we can certainly pose restrictions on the DL systems limiting them to operate only in the realm where every decision made and understanding acquired is fully understandable to us. While such an approach may let us sleep in peace, it may restrict most of the DL system's potential, allowing us to only partially enjoy its benefits.

Until few years back the dilemma discussed above was only a thought experiment. But as the AI, ML, DL technologies gather speed, this issue has started showing its head in real world situations. As a counterpoint to the Syracuse example, we can look at another example where the results may not have been equally beneficial. Recently the University of Pittsburgh medical center ran a project. The idea was to leverage the capabilities of a neural network to correctly predict whether the health of pneumonia patients that show up in the hospital is going to get worse or better. If the patient is not expected to get worse, then the cost of treatment can be optimized, and the patient can be sent home with some medication. Otherwise, patients could be admitted into the hospital (incurring higher costs) to monitor them closely (thus justifying the cost).

Using the data provided, the program was indeed assessing the outcomes correctly, thus helping the hospital considerably, and easing admission congestions. But actual doctors who were analyzing the results noticed a disastrous trend. Anytime a patient with asthma showed up with a case of pneumonia, the DL program tended to predict/would predict that they will get better soon and so can be discharged right away. Physicians knew very well that asthma patients with pneumonia have a very high probability of landing in a worse health condition and so the usual practice is to jump the line and admit them into the ICU. Since that has been the practice, those patients' condition usually never got worse. But the DL program using the data set it received, deduced that the asthma patients with pneumonia never got worse, without realizing that this was *because* of the extreme levels of care they received in the ICU. Since that detail wasn't available on the dataset it used, it didn't have that knowledge, though for physicians in the hospital this is an obvious truth.

In statistics classes there used to be a simpler example discussed to explain this phenomenon. If you go strictly by the available data, you will see a clear correlation between ice cream consumption and death by drowning in the USA. Graphs will clearly show drowning deaths increasing as the ice cream consumption increases and decreasing as the consumption decreases. Based on that statistics, one might be tempted to ban ice cream consumption to prevent drowning deaths. But when we step back and take a better look at the big picture, it will become obvious that people tend to eat more ice cream in summer, and summer is when they also swim more in swimming pools. So, the increase in drowning deaths is due to increased rates of swimming and pool usage and not due to ice cream consumption.

In statistics classes, teachers reiterate ad nauseam the fact that two measures going up and down together does not imply that one is causing the other, as in the saying, "Correlation is not causation." While the ice cream example may sound like a silly one, even experts often stumble into this pitfall and reach erroneous conclusions while handling complicated data sets, despite having learned this lesson. When the data set used in the analysis is incomplete, you can't blame the DL algorithm for reaching the wrong conclusion while trying to figure out whether the patient will get worse or better. But for physicians working in the hospital, figuring out that pneumonia is dangerous to asthma patients is a no-brainer. If we ask whether a computer can ever learn such facts and correct itself, the answer is yes. When it continues to send such patients home initially and they continue to die, that increase in death rate will allow the algorithm to correct itself. But who is responsible for all the deaths that would have to occur for the algorithm to learn this truth?

Pondering over this problem, we may conclude that if DL algorithms ever lead to actual deaths, that means they are at a very primitive stage and so should not be employed in real life situations for a long time. But the same DL techniques have successfully predicted where the next flu epidemic is expected to peak, allowing healthcare workers to ship flu vaccines and medicines to those locations ahead of time, thus preventing the epidemic completely. Factors as varied as weather reports, background information on the local public (do they shake hands when they meet and greet or just do a "Namaste" as people in India traditionally do?), the incidences of flu in previous years, newspaper reports on child births, sports events and so many other datasets have factored into that prediction, which would be hard for human beings to use in their analysis.

Though we are not able to understand individual steps involved in how the algorithm came to its flu prediction conclusion, by listening to its conclusions we are able to ship needed medicine to that area and prevent deaths. So, if we push such technologies aside too sternly, we may end up not saving lives that could potentially be saved. So, what is the correct approach? Is it presuming that computers will become smarter than us at some point, such progress cannot be stopped by us and so we need to embrace such progress to take full advantage of its capabilities, even if we don't fully understand the process? Or is it deciding to prevent anything that we can't parse and fully understand on our own, within the limits of human capabilities?

China Brain instead of Chinese Room

One school of neurologists thinks that functioning of the human brain is nothing but neurons in our brain triggering on or off forming various neural networking patterns as they respond to various scenarios or incoming input. In the Chinese Room thought experiment proposed by John Searle, we are analyzing the functioning of one human brain. Instead, let us take up a different thought experiment. As per this experiment, presume there are one billion neurons in a normal brain. Now let us say the feeling of sadness is felt when a subset of 600 million neurons are triggered in a specific sequence. So, starting from one or a small set of neurons, when these 600 million neurons are triggered, they wake up and form a network pattern that makes the human being carrying those neurons feel sad.

Now considering each Chinese citizen as one neuron and the entire population of one billion people in China forming the equivalent of one human brain, let us create a process by which one or a small set of Chinese people start contacting a total of 600 million fellow citizens via telephone in the same sequence as the neurons do. When the phone rings, the person answering may not even need to speak anything into the phone. They just pick up the receiver and then dial others they are supposed to contact. That's it. Thus, we trigger the same 600 million subset of Chinese people to wake up and connect via telephone as the 600 million neurons do.

If we think of each individual neuron as a cog in a big machine's internal wheels, each individual neuron may not understand what the total brain is feeling/thinking just like individual cogs in the machine may not have any understanding of what the total machine is doing. In the same manner, the individual Chinese citizen receiving and placing phone calls may not understand what the total setup is trying to do or moving towards. But just as a human brain will feel sad when

Ned Block

the 600 million neurons have fired in the correct order, can we presume the total nation of China is in a sad state when all 600 million Chinese citizens in the aforementioned process have been called and connected?

New York University's Ned Block proposed this thought experiment circa 1978. Through this experiment we can analyze the idea of emergent properties, where parts of a bigger system may not understand the whole scenario fully, while the system as a whole, can develop new types of awareness. We can also use it to argue that human consciousness and self-awareness are all an illusion. Those who are familiar with the Hindu religion may recall that it espouses the idea that our birth, life, death on this planet are all an illusion, referred to as Maya. So, we should not get attached to material possessions or what happens on a daily basis in life, but elevate ourselves to higher spiritual planes where these illusions no longer bother us. It is amusing to see a resemblance or supporting evidence to that Eastern theological notion in Computer Science oriented thought experiments.

--------- o0o ----------

Enhancing Comprehension

The eye sees only what the mind is prepared to comprehend.

— Robertson Davies

Newcomb Paradox

You are invited to be contestant on a TV game show, something akin to Who Wants to be a Millionaire? You don't need to answer silly questions like how many legs does a spider have, etc. to get through initial rounds and kill time. In this strange game, there will be two boxes placed on the stage. Since box 1 is made of glass, you can clearly see there is $1000 inside. But box 2 is made of steel and sealed, preventing you from knowing what is inside. As per the rules of the game, box 2 either has a million dollars inside or has nothing. The entire game is simply a selection you are going to make. You can either choose just box 1 or just box 2 or choose both. Your goal is to collect the maximum money possible.

The funny thing about this show is the presence of banker character like the one you might have seen in the TV show "Deal or No Deal". This banker's specialty is his proclaimed clairvoyance. A week before you left your home to come to the TV studio to participate in the game, this strange banker is able to guess what is going to be your eventual box selection. If the banker thinks you are going to choose only box 2, he is going to place a million dollars inside that box and seal it. Alternatively, if he thinks you are going to choose both boxes, he is going to leave box 2 empty and seal it. It is guaranteed that he seals the box one week before you left home. When you come to the studio and play this game a week later, will you pick up just box 2 or both boxes?

If we think about this choice carefully, we will realize that the decision is not so simple to make. Since your goal is to win as much money as possible, there is no point in choosing box 1 alone. So, depending upon what decision you make, the game branches into four possibilities.

Possibility 1: The banker predicts that you will choose box 2 and so places a million dollars inside. You also choose box 2 and you end up with one million dollars.

Possibility 2: The banker predicts that you will choose box 2 and so places a million dollars inside. But you cleverly choose both boxes and so you end up with one million and one thousand dollars. This is the maximum money you can acquire.

Possibility 3: The banker predicts that you will choose both boxes and so leaves box 2 empty. You do choose both boxes and so you end up with just one thousand dollars.

Possibility 4: The banker predicts that you will choose both boxes and so leaves box 2 empty. You decide to pick only box 2 and so end up with nothing.

This thought experiment proposed by the University of California Physics Professor William Newcomb is often referred to as the Newcomb Paradox. Though it was created by Newcomb, Robert Nozick is the one who wrote about it making it famous. As it became part of the pop culture, a few wondered aloud as to what the paradox here is. It is actually lurking in the strategy people propose to score the maximum amount of money while playing the game. Half the people trying to come up with a strategy fall in the 'single box' camp. They are very clear that choosing box 2 is the correct approach and so one million dollars is the maximum amount of money anyone can win. The other half belong to the "double box" camp. They are convinced that since the boxes have been sealed a week back, their current decision cannot change what is inside and so the player must pick up both boxes to maximize the winning. They argue that if the banker had predicted that we will take only box 2, we get lucky as we will end up with one million and one thousand dollars. If not, we will get the thousand dollars in box 1. Since the money inside box 2 is not going to change at the instant of selection, they don't see any reason to leave a box out. They do ask a very interesting question. Just for an argument, let us say you are the player in the game and someone you trust is sitting on the other side of the boxes. Now let us say the second box's one side alone is made of glass making its inside contents visible to your trusted partner, who can't tell you what is inside. Once they find out whether there is a million dollars inside or not, what would they want you to do? Choose just box 2 or both boxes?

Robert Nozick

In general, people on both camps are surprised and puzzled that the other side is not about to understand that their logic is incorrect. Each side gets irritated that while the answer is obvious, somehow the other team is not able to figure out this simple strategy. Newcomb himself fell into the single box camp. But Nozick was initially a double box camper. But after reading multiple articles and arguments related to this paradox, he subsequently equivocated a bit claiming there is logic on both sides. Explain this problem to your friends/family, make sure they understand it well and then ask for their strategy and try tabulating them. You might be surprised by the results.

This paradox has driven many philosophers and game theorists crazy for decades. For now, just remember the main points in this thought experiment. After we discuss a few other TEs, we will see if we can connect those TEs to this paradox.

Shadows in the Cave

Greek philosopher Plato who lived in Athens circa 400 BC proposed a thought experiment. His goal was to push his followers not to believe what you hear or see but trust things only after you have investigated, analyzed and understood something in its entirety. This experiment presented in his book called *The Republic* is structured as if his teacher Socrates is teaching this idea to Plato's brother.

Plato

There is a dark cave. There are several prisoners imprisoned inside with their hands hand-cuffed permanently to the wall. All they can do is simply stare at a blank wall in front of them. Since they have been imprisoned right from the time of birth, they have never been outside and so have no idea of how the outside world looks. Behind them is a torch that has been burning forever.

There is a wall in between the torch and the prisoners. There are a bunch of puppeteers carrying human and animal looking puppets mounted on sticks. As they walk around, the puppets cast shadows on the wall that the prisoners have been looking at permanently. Since the prisoners have never seen the actual puppeteers, as far as they are concerned, whatever shadow they see on the wall in front of them is their reality. They even mistake the sound of the footsteps and voices caused by the puppeteers as coming from the shadows on the wall. Within the society of those prisoners, anyone who can talk in detail about the shadows is considered the leader and an intellectual.

Now let us say some of the chained prisoners are freed. First when they turn around and see the torch that is throwing the light along with the puppets and the puppeteers, they realize that what they thought was reality has no relation whatsoever to the real world out there. When they eventually leave the cave and step outside, their eyes have a hard time getting adjusted to the bright light outside as they have always been in the dark all their lives. Once they get used to the light, they will come to see and appreciate sun, moon, rainbow, lake, ocean, fields and so on. They will then be able to feel sorry for the prisoners in the cave who are totally uninformed and so are completely mistaken about their perception of the world. Plato posits that just like the prisoners had "knowledge" of the entire world formed by viewing the shadows in front of them, we have "knowledge" of the entire world through the information flowing in through our five senses. Enlightened philosophers and intellectuals amongst us are the equivalent of the prisoners who have been freed. They have moved further steps away and ahead of us and have started learning the details of the world in a much better fashion. Plato argues that instead of simply pursuing this new knowledge endlessly, they also have a moral obligation to return to the cave and teach us, the prisoners, the truth.

In today's world, people who simply believe what they read in print or on echo chamber websites could be considered the equivalent of the chained prisoners in the cave.

Mary's Room

Frank Jackson approached the same issue of perceiving the world from a different angle. He is a contemporary Australian professor of philosophy. The protagonist in his story is a brilliant neuroscientist named Mary.

Frank Jackson

As per his thought experiment, she has been living in a room that has only black and white colors since her birth. All her research takes place using a black & white computer terminal. She is an expert in neuroscience and so understands how the human brain works perfectly. Exposed to sound or light, colors or feelings, fear or joy or whatever other human experiences are possible, she knows how each neuron in the human brain will perceive and react to it. Thus for example, when blue light falls on our retina, what happens in the eye, how the brain registers that information, when we say "I see blue light", what part of the brain creates that message, how that information reaches our vocal cords and how that message gets delivered externally are all fully understood by her. At this point, though she intellectually understands all that happens when a human being perceives blue light, one day if she really leaves her black and white room to step out into the real world to see real blue sky for the first time in her life, will she experience a new understanding of what blue light is? If she will, then understanding becomes something occurring beyond simple physical, chemical, or biological reaction taking place in the realm of self-awareness and consciousness. So, Jackson says that the notion of understanding has an innate dualism in it.

You might have heard a Shakespearean quote claiming that the whole world is a stage and we are all just actors on that stage playing a role assigned to us. In the subsequent chapters, we will see all the connections intertwined between these ideas embedding one perception into another.

--------- o0o ----------

Thoughts to Spoil Your Sleep

It is far harder to kill a phantom than a reality.

— *Virginia Woolf*

I am planning to wander through few confusing and weird corners. Let us see if we return to the base safely.

Brain in a Vat

Proponents of Solipsism claim that all we can know is only our individual consciousness. If we don't have even that, we cannot even get into this discussion. So, that is an immutable truth. But beyond that little limit, starting from anyone we are interacting with, all the way to the entire universe, we can never know what is real or claim with certainty what is true. Everything we perceive could be an illusion. Many think that the seventeenth century French philosopher Rene Descartes is a Solipsist. Some do dispute that. But his signature line "I think, therefore I am" that we all have heard of, seems to imply that he could belong to this camp. With the presumption that only that line is a knowable truth, he set out to develop and prove the complete framework required for human beings and societies to function. We are not sure if he got very far. But while searching for those evidences, he popped a question. We perceive and understand the entire universe using the five senses we possess and the information they send into our brain. When that is the case, how do we know that there is no ghost or demon that has

Hillary Putnam

taken over our five senses sending in messages though our hijacked interfaces, making us falsely believe that we live and breathe in this world?

An extension of that question is the Brain in the Vat thought experiment. This was popularized by contemporary American philosopher Hillary Putnam. Before brushing it aside as a silly idea, let us dive in to see what its implications will be in case it is true. Let us say a mischievous scientist sedates you completely while you are asleep, takes out your brain and sets it up in a special fluid that will allow it to function normally. He then attaches a set of electrodes to your floating brain and connects them to a powerful computer. The computer is capable of feeding in all the input your five senses will normally send to your brain via your nose, eyes, ears, mouth, skin, and so on. Since the scientist programs the computer to send in whatever signals he wants, he can make your brain think you are walking on the moon, swimming inside an ocean or simply sitting at your home drinking coffee.

You may be ready to laugh it off saying neuroscience, computer technology or various other fields that need to come together to make this happen are not anywhere close to such sophistication. But if I tell you that your knowledge allowing you to make that statement, this book you are reading are all inside this virtual world lingering only inside your floating brain manipulated by the scientist, how can you disprove that statement? You can see this concept leveraged in the movie The Matrix, several Star Trek TV show episodes as well as several sci-fi novels.

Now let us revisit the Newcomb Paradox we encountered in the last chapter. (If you haven't read that section and skipped to this one, it might be better to read that first to get the full context.) We are meeting that banker character again. For all you know, his never-failing clairvoyance might be based on the results of the experiments his mischievous scientist friend has been conducting. To determine whether you will pick up just box 2 or both boxes in the real world, he might have made a copy of your brain or might have picked the brain of an identical twin of yours. Placing that brain in the vat he will then feed the needed input to make it play the game. The immersive sensory input provided to that floating brain will make it believe that it is traveling to the game show, checking in, playing the game in reality and is making the choice of picking just box 2 or picking both boxes. Whatever choice that brain makes in the simulation is what the banker says he figured out through his clairvoyance that you will make when you play the game, claiming that he never fails. If you cannot conclusively prove that you are not a brain in a vat, there is a possibility that what you think you are doing in the real world could be a simulation run by a banker and his scientist friend who are the ones in the true real world.

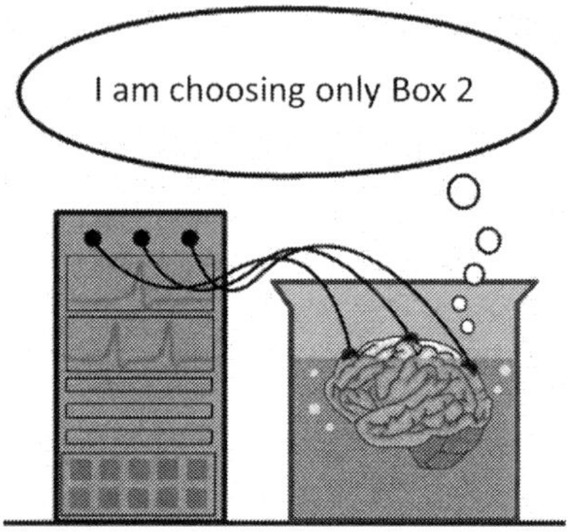

So how should you play the Newcomb Paradox game if you might be a brain in a vat? Suppose you belong to the single box camp, so you think that choosing just box 2 is the clever choice. Then your simulated brain would pick just box 2. Based on the simulation of that decision, the banker in the real world would place a million dollars in box 2 and seal it. The real you in the real world may be able to pick just box 2 or both boxes, but as long as you are not sure if you are in a simulation or in a real world, you should always pick just box 2. Otherwise, the probability of you returning home with just one thousand dollars increases considerably. This is the argument of the single box campers. If in a parallel world, such brain in a vat simulations are possible, perhaps it might be possible that what you decide now, determines what gets put into box 2 one week back. So, what is happening in a future point in time affects what happens in the past. If you don't think this makes any sense, you can skip this paragraph and move to the next. If you find this line of thinking intriguing, you can read more about Timeless Decision Theory on the web and lose even more sleep. If you are not confused till now, we can look at what Putnam himself thought of this Newcomb Paradox thought experiment. He said if the brain making the decision at a given point believes that everything is taking place in reality, the decisions being made by that brain being exactly in line with what will happen in the real world is what makes logical sense. If you accept this logic, it means we can pick the boxes as we like without worrying about a simulated world or us being in a simulation. Now we can step up one more level and have even more fun.

Singularity?

Earlier we saw that AI technology is growing by leaps and bounds each year. As this progress continues, we may see two possible tipping points in the near future.

1. Computers passing the Turing Test discussed in the third chapter, without resorting to programming trickery as currently being done, could be one such tipping point.

2. As the power of computers continues to increase, a lot of effort is simultaneously going into AI to leverage this increase in power to improve the intelligence of computers. As a result, AI can soon reach a point where, without human aid and intervention, computers can continue to learn on their own and decide what they

want to learn. When computers reach that point, the normal guard rails governing human behavior will not restrict computers from what they want to do. For example, unlike human beings who are encumbered by biological limitations, the reach of computers in accessing large volumes of data and processing capacity grow at extraordinary rates. Computers can use that growth to learn faster and faster. Since there is no obvious end to this acceleration within the horizons visible to us, it is very hard for human beings to assess where this will lead us in the future. AI experts refer to this state as Technological Singularity.

This term, 'Singularity', introduced in 1958 foreshadowed a potentially confusing future slowly approaching us. The growth of human knowledge and the growth of computers' knowledge via AI are running forward as two parallel lines for now right in front of us. In the future, these two lines may merge into a single point. From that time onwards, the knowledge of computers will start to exceed human knowledge. It may never be possible for human being to catch up due to the inherent limitations of biological beings as to how fast they can learn and grow. The idea of Singularity suggests that once that point is reached, it will be hard to fathom how computers will treat human beings (or any other life form on earth), how our world will look, and how our laws & ethos will change.

How do we ensure that computers will remain our friends and won't turn into our enemies is a big worry amongst very many organizations and intellectuals pondering over these issues. In one of these debates, someone with a pseudonym Roko introduced a thought experiment that went viral in these communities a few years back. There is also an interesting connection between this thought experiment and the Newcomb Paradox. We will get to it before the end of this chapter.

Roko's Basilisk

A basilisk is a cross between a snake and dragon found in European mythology. This beast considered the king of snakes can kill anyone by simply looking at them. The myth builds up to claim that there is no equal to its terror. The TE introduced by Mr. Roko apparently destroys those who learn about it, just like the mythological Basilisk and so the TE is named Roko's Basilisk. So, beware and consider yourself duly warned before you read further!

We said earlier that once computers reach the point of technological singularity, we don't know how AI will grow beyond that point and what it will morph into moving forward. While AI by itself is racing ahead to achieve its new milestones and targets, it may destroy anything in its way that is slowing down its progress. Thus, for example, if it wants to construct a new biological computer for which human beings provide the needed raw material, it may turn out to be a disaster for our kind. It may not mean that AI likes or hates human beings. Just like we, human beings, use up and destroy animals or things we find on earth to reach our goals without really worrying about how the animals or things will feel, AI may not care as to what we may think or feel about us being used as raw material for some new biological computer. Since the Basilisk is very intelligent, it may reach back in time from some future beyond singularity and grab us by our necks forcing us to do something to help its cause. This was the core idea suggested by Roko.

Just like Newcomb paradox, this Basilisk gives us two choices and forces us to make a choice. Basilisk's preference is you opting to take box 1 (or both boxes). If you do take box 1, you affirm that Basilisk is a true manifestation and you agree to support its causes. (Alternatively, if you choose both boxes, the effect is the same since box 2 is empty.) That is the end of the story.

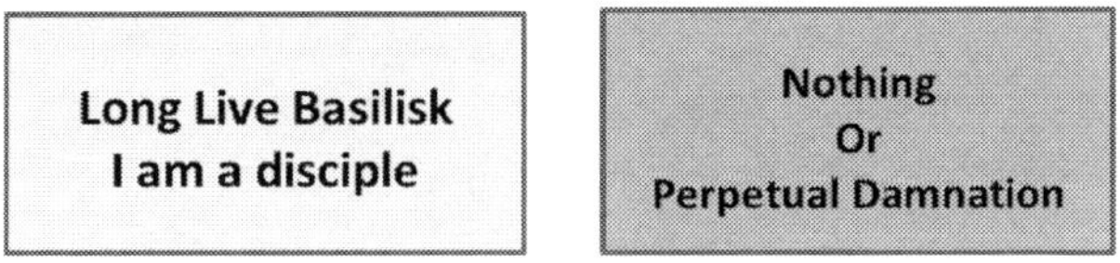

Box 1 Box 2

Instead, if you think that this is all total nonsense, you will choose box 2 alone, expecting it to be empty. Since Basilisk knows your choice before the boxes were sealed, it would have then cursed you to eternal damnation. This is the thesis of this philosophy. You can push back saying how can a future occurrence that is going to take place only after the singularity

Eliezer Yudkowsky

has been reached affect box 2's contents now?

You are free to brush all this aside saying this is some confusing sci-fi nonsense. People like Eliezer Yudkowsky think that these are not some drug induced wild hallucinations but a reality that is bound to roll out if you extend the growth of AI to its logical conclusion. So, Yudkowsky and those of his ilk are trying to figure out how to make sure that the growth of future AI remains human friendly by arranging large conferences under this umbrella topic. Even if you try to guide the growth of AI in a direction that will support human civilization, they think that there is a very high probability that things will go haywire leading to unpleasant consequences for our race. As per their analysis, if you delve into this line of thinking and wonder about the Basilisk a lot, automatically the necessity for you to become its slave increases. If you refuse to become a slave, the probability of you having to endure its torture increases. So, they recommend you walking away from this topic respecting this warning. Alternatively, if you are laughing now hysterically at this notion and want to confront the Basilisk yourself, you can logon to LessWrong.com like websites, and try battling topics like Coherent Extrapolated Volition (CEV). While you are doing that if your neighbor complains to you that someone from the future is controlling and torturing them, please let me know.

Quitting all these sleep depriving TEs, we will jump into the world of Mathematics related TEs in the next chapter. I hope you don't feel drowsy as soon as we start talking about Math TEs.

Note:

* On a December 31st Rene Descartes's home is hosting a New Year's Eve party. Mrs. Descartes has prepared a bunch of nice, luscious looking cakes to be served after the New Year arrives. Rene Descartes is sitting next to them. When some of the guests try to grab a piece even before mid-night, he panics and shouts, "I think they are for 1 am". Guests who misheard him converted that statement into a big tenet in existential philosophy!

--------- o0o ----------

Mathematical Thought Experiments

Without mathematics, there's nothing you can do. Everything around you is mathematics. Everything around you is numbers.

— *Shakuntala Devi*

Hotel Hilbert

David Hilbert

In middle school we learned that you can begin with 1, 2, 3 and keep counting forever. The series will never end and will lead you to the biggest possible number which is infinity. As you may remember, that infinity is a peculiar number. You can't ask if it is an odd number or an even number. Now that we have said it is the biggest number out there, is there anything bigger than that? To answer questions of this kind, a famous 20th century German mathematician named David Hilbert prescribed a thought experiment. George Gamow popularized that TE couple of decades later circa 1947 in his best seller book One, Two, Three...Infinity.

In his TE, Hilbert constructs an imaginary hotel that has an infinite number of rooms. Even the hotel's advertisement tag line reads, "Who said no vacancy?" The management guarantees that even if they are full, they will never turn away a guest and will always find a room for them. Let's say there is a huge conference on Mathematics taking place in town that has filled all the hotel rooms including the Hilbert Hotel. On that evening a new VIP shows up demanding a room for him to stay that night. Hotel is fully occupied. So, what can they do? The hotel manager thinks for a minute and tells the VIP that there is a room for him and asks him to wait just few minutes. He then asks all the guests in the hotel to look up their hotel room number, add 1 to it and move to that new room number. In other words, they are all asked to move to the next room. Thus, since the guest in room 1 moved to room 2 and guest in room 2 moved to room 3 and so on, room 1 now becomes vacant. Manager assigns that room to the VIP waiting at the counter. What if a new VVIP shows up next? No problem. We just need to ask the guests to repeat the same process that will put the VIP in room 2, resulting in room 1 being vacant again to accommodate our new VVIP. Thus, irrespective of the number of new guests that show up, we can repeat this process creating new vacancies and accommodate the new guests. Since Hotel Hilbert has infinite number of rooms, this is always possible.

When some experts heard this method of allocating rooms to new guests, they pointed out that when the hotel was full initially, all the guests were settled in their room resting. When you ask everyone to move to the next room, there is at least one person constantly in the hotel corridor trying to relocate to the next room. As per their argument, the person walking in the corridor to move to the next room doesn't have a room at that instant. Viewing the same event from a different perspective, this is equivalent to asking the new guest to take up room number infinity. It will take him forever to reach that room and so he will keep walking forever. So, in their view, this cannot be seen as a real room being allocated to him. But if you follow Hilbert's algorithm of assigning rooms, each guest will find their room within a finite amount of time, settle down and start resting again. The Hilbert camp's response was, "The experts who push back didn't really understand the algorithm correctly".

Now let us say a competing Computer Science organization also decides to host a conference in the same town. A large number of participants for that CS conference show up in an unusual bus. We are calling that bus unusual since it has infinite seating capacity. In other words, that bus is bringing in number of guests equal to infinity to the town, and they all also need to be provided hotel rooms. We don't need to worry as to whether such a hotel or bus is possible in this world. Isn't that the main convenience in running thought experiments?

Since Hotel Hilbert is the most famous lodge in town, bus heads to it right away. Now the hotel manager must provide rooms for an additional infinite number of guests again. After thinking through it for a minute, he comes up with a plan and declares all the new guests will be provided rooms.

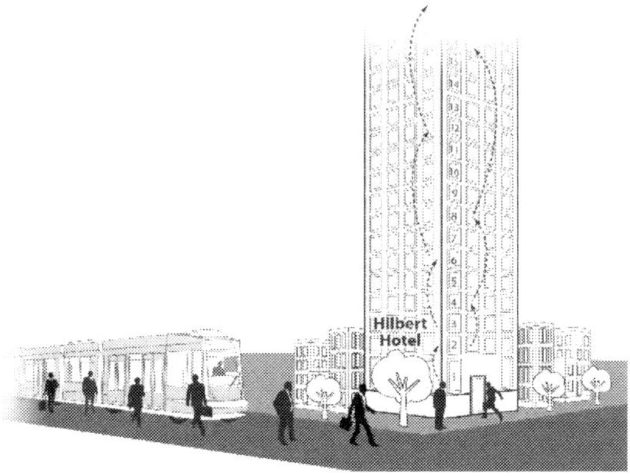

He tells all his existing hotel room guests to find out their current room number, multiply it by two and move to the room with the resulting number on its door. So, the guest in room 1 moves to room 2, guest in room 2 moves to 4, guest in room 3 moves to room 6, guest in room 4 moves to room 8 and so on as shown in the figure. Now all the odd numbers rooms become vacant. Since the number of odds by itself is also equal to infinity, every guest in the bus with infinity seats could now be accommodated in the hotel.

Let us see if that brilliant manager can handle a situation that is an order of magnitude more complex. To create this situation, instead of just Math and CS organizations, let us presume that every organization you can imagine decides to run a conference, bussing in infinite number of attendees each, to this same town. Since infinite organizations bring in an infinite number of people each, this will result in an infinity squared number of guests showing up asking for rooms in the hotel. The manager surprisingly claims that he can provide rooms to everyone in that infinity squared crowd. How?

If his plan is to provide rooms for all the people who came in bus 1 first and then start working on guests who arrive on bus 2, then he may never get to the first person on bus 2 since bus 1 itself has an infinite number of guests. In other words, no one who came in on bus 2 onwards will ever get their rooms. So, how does he manage this situation? Answer lies in the sequence shown in the picture above as he starts allocating rooms. He assigns the first available room to the first person on the first bus. Then assigns the next available room to the second person that arrived on bus 1. After this point, instead of moving on to the third person on bus 1, he assigns the third available room to the first person who arrived in bus 2. He then assigns the fourth available room to the first person that arrived on bus 3 as shown. By following this sequence, the manger claims that he will be able to assign rooms to each person in the crowd a room in a finite amount of time, without having to make anyone wait an infinite amount of time. Mathematician George Cantor said, while all this sounds good at first, there is a problem in this algorithm. Cantor is Hilbert's senior. Though the proof he provided to show the issue in this approach is more than hundred years old, its elegance still remains surprisingly beautiful. Cantor said that the number of fractions in between zero and one alone is an order of magnitude bigger infinity. So, there aren't enough numbers in the first order infinity defined by the natural number set of 1, 2, 3, to map all the fractions in between zero and 1, as it is a much bigger infinity. This was initially confusing to many. Didn't we say infinity is the biggest number out there? Then how can there be a bigger infinity that is bigger than the original infinity we were talking about? Cantor provided a simple but elegant proof to prove his point. Let us map it to the Hotel Hilbert room allocation process to understand how it works.

Georg Cantor

We said Hilbert Hotel has rooms numbered as 1, 2, 3 running all the way to infinity. Now let us say one of the conferences taking place is investigating the number of fractions that exist between digits zero and 1 and there is one attendee that represents each one of those fractions between zero and 1. If we try to accommodate all these guests into Hilbert Hotel with infinite rooms, you still won't have enough rooms. Even if you totally empty the entire hotel and try to allocate rooms for guests who came in for this conference alone, we will not have enough rooms. Many mathematicians refused to accept that there can be smaller and bigger infinities. But Hilbert jumped up in joy claiming no one can force us out of this new paradise Cantor has created for us. Perhaps Hilbert's happiness is the one that defines the actual infinity or perhaps it is so large that it is undefined.

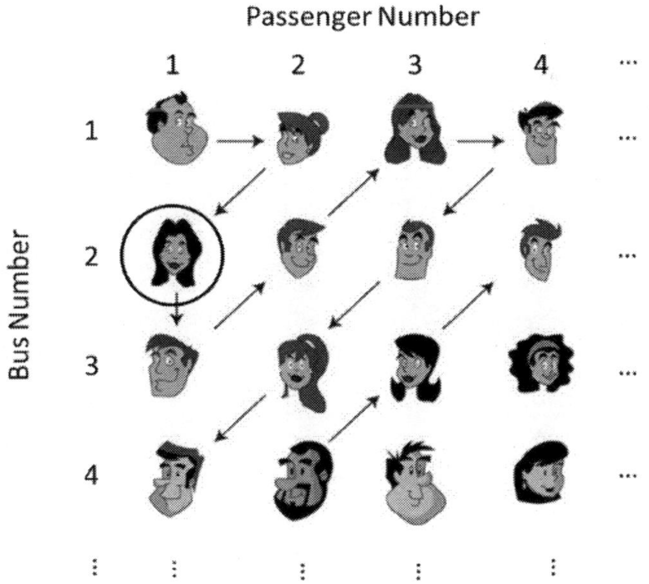

We can explain Cantor's proof this way. Looking at the picture above, we easily understand that both rows and columns of that figure extends to infinity as there are infinite passengers in each bus (represented by one row each) and there are infinite buses (represented by one column each). If we divide the number of the passenger by the assigned number of their bus, we will get a fraction. For example, the first passenger (1) who came in on the second bus (2) is that lady with a circle around her. The fraction that maps to her will be ½ or 0.5.

Room Number	Passenger Number	Bus Number	Passenger's Fraction
1	1	1	1.000000000000000000000000000...
2	2	1	2.000000000000000000000000000...
3	1	2	0.500000000000000000000000000...
4	1	3	0.333333333333333333333333333...
...	...	...	...

The hotel manager is assigning a specific room represented by a specific natural number (i.e. room number) on it for each guest. So, he is actually mapping each such fraction to a natural number. Since she is getting room number 3, the fraction ½ gets mapped to natural number 3.

Since the rows of this setup (number of buses) and columns (number of passengers in each bus) are each infinite, we can see infinite number of such ratios being generated and then getting mapped to one natural number precisely following the room allocation algorithm. (Though the terms fractions and ratios may appear interchangeable initially, fractions are rational numbers whose decimal expansion must either end or repeat but can't continue endlessly. The decimal expansion of an irrational number continues without repeating, thus making π a well-known irrational number. We will use the term ratios to talk about these numbers just for convenience, though it may not be the correct mathematical term.) Since rows and columns are infinite, this process will also capture every imaginable ratio of a/b where a is one natural number that forms the numerator and b is another natural number that forms the denominator of the ratio. Cantor took this list of ratios (think of it as an infinitely long list with each row containing one ratio that has infinite decimals) and created a brand-new number by picking one decimal that differs from each row.

Let's say these are the list of ratios he is starting with:

0.**4**93845093485093…

0.2**7**5263452346543…

0.76**8**762348762374…

0.824**5**25984759879…

…

He forms a new ratio by picking all the bold-faced digits in each one of these numbers. This will be

0.4785..

He then changes each one of these digits to some other random digit in the range of 0 to 9. Let's say this new fraction happens to be

0.2139..

It is easy to see that this new number will differ from the list of ratios we started with at the beginning at the first decimal (2 differs from 4). It will differ from the next ratio in the list at the next decimal (1 differs from 7). It differs from the third ratio at the third decimal (3 differs from 8) and so on. If we think for a minute, we will realize that this newly created number is still between zero and 1 but differs from every ratio that is in the table in at least one position. In other words, this new number will not be found in the list of ratios created. This means there is a valid guest (since the newly created ratio is a valid one between zero and 1) but won't get a room in the hotel (since this is not mapping to any natural number).

Thus, by showing that this new ratio has escaped the mapping process, Cantor showed that there will always be more ratios unaccounted when you map them to natural numbers. In other words, this proves that the number of ratios between zero and 1 is more than the natural numbers.

Achilles and the Tortoise

Greek Philosopher Zeno of Elia who lived 2,500 years ago, suggested that even movement is an illusion. To prove his point, he proposed a thought experiment. Achilles was a world-renowned athlete and a marathon runner in those days. In

Zeno's TE, Achilles is forced to compete in a 100m running race with a tortoise. Achilles thinks this is a silly race and so tells the tortoise, "Me competing with you is not really fair to you. So, let me give you a 50m head start. Until you reach the halfway mark, I won't even start. So, go ahead. Once you reach the 50m mark, I will start and will still easily win this race". Tortoise agrees and starts off. When it crosses the 50m mark, Achilles starts running towards the finish line. Zeno asks us to determine who will win that race.

As per his logic, though Achilles is a much faster runner than the tortoise, during the time he started and reached the 50m mark, the tortoise would have moved a little further away from the 50th meter mark. Just to make the discussion easier, let us say this turbo charged tortoise runs at 20% speed compared to Achilles. So, by the time Achilles reaches the 50th meter mark, the tortoise would have reached the 60th meter mark (i.e. an additional 20% of 50m, which is 10m). When Achilles runs this additional 10m distance, the tortoise would have moved an additional 2m. When he covers this extra 2m, tortoise would have moved yet another 0.4m. So, as Achilles catches up to each point on the track, tortoise would have moved some additional distance. So, Zeno argued that Achilles can never go past the tortoise to win the race! What is wrong with his logic?

You may say thinking through this problem gives you headache, all you know is by the time he reaches 100m, Achilles will be ahead of the tortoise and so I should leave you alone. But then you had to concede that while you are getting away, you haven't pointed out any real flaw in Zeno's argument. Then how can we explain this paradox?

Mathematics has a lot of continuing series that never end. A very simple example is 1 + 2 + 3 + 4 + 5 + … series. We all know that there is no end to this series as it races towards infinity. The sum of this series will be infinity. That's all we know. A series of this kind where the sum of the series becomes substantially larger and larger in each step without any end in sight is called diverging series.

Now consider a different series ½ + ¼ + 1/8 + 1/16 + 1/32 + … Obviously this is also never ending. But inspecting just for a minute will help us easily conclude that while the series is never ending, its eventual summation is going to be 1. Such a series is called a converging series. Thus we see one type of series where each step leads to a larger sum that keeps growing enormously as you continue walking through the series and another type of series where the sum seems to converge to a specific value as you move further into the series elements. If a series is converging, there are mathematical formulas that will help us determine the value to which the series will converge. Since Achilles seems to be running inside such a converging series, in a time window we can precisely measure, we can be sure that he will reach the end of the series. So, we can safely conclude that he will win the race against the tortoise.

Perhaps credit card bills and EMI (Equal Monthly Installment) plans are all part of such a never-ending series of illusions. That is why people who enter those illusions often never manage to successfully navigate through the series and emerge on the other side but end up paying interest forever to an ever-elusive principle that always seems to be a little further ahead for one to catch up.

--------- o0o ----------

Impacting Feminism

A feminist is anyone who recognizes the equality and full humanity of women and men."

— *Gloria Steinem*

Let's use thought experiments in this chapter to explore gender bias that works against women.

In 2014 there were media reports about a private company advertising to recruit volunteers to go to planet Mars on a one-way ticket. They apparently received about 200,000 applications. After running several tests, that company narrowed down the list to about 700 candidates. Their plan was to select just a dozen candidates deemed suitable for the journey after several more rounds of testing and filtering process. The candidates may finally be divided into small teams that are expected to travel together or as individuals fit for the trip. NASA, ESA, ISRO, and other such well recognized space research institutions have not yet started developing rockets declaring their intent to ship off dozens of travelers to Mars. No one has promised funds to put such a massive project on track either. But the non-profit institution called Mars-One running these selection processes claims that these activities currently taking place will trigger public and private enthusiasm for such a project resulting in needed funding and support. We will see if this dream comes to fruition in our lifetime.

Irrespective of whether that plan comes to pass or not, a woman named Stacy Ritz proposed a thought experiment based on shipping people to Mars. In the initial years of the 22nd century, the UN decides to colonize other planets to find homes for the growing population of human beings on planet Earth. To plan out the migration in a scientifically sound manner, the organization decides to send explorers first. It runs tests all over the world and selects John from Finland and Joy from Thailand. They are placed on two different spaceships and rocketed off to Mars and Venus. The futuristic spaceship is equipped with enough oxygen, water, food, as well as special equipment that can clone human beings. After their long journey, John and Joy reach Mars and Venus respectively and set up their labs. In subsequent months and years, they clone themselves to create a small population to assess the conduciveness of the planet for large scale human migration. The growing clone population in each planet lives comfortably for a whole century.

In the following century, as per the original plan, a team of scientists embark on an interplanetary trip to assess living conditions on both planets. To determine if these two planets are appropriate for long term human colonization, scientists plan on running a battery of tests on the residents of both planets.

1. Measurement of their LDL cholesterol level.
2. Assessment of their language skills.
3. Determination of the efficacy of various medications.
4. Physical stamina, strength of their legs
5. Population's height

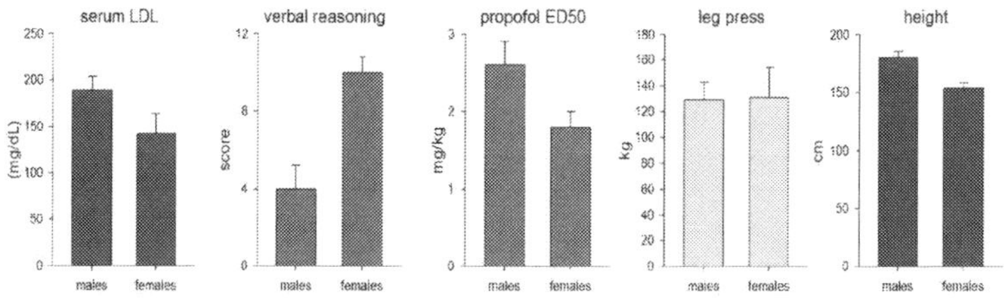

After gathering all the required data, they return to planet Earth, analyze the information and publish a report containing the graphs shown above. Though they live in two different planets, innate differences between men and women are perceived to be reflected in these results. Journalists and public analyze and consume the results.

The way experiments were conducted, and the instruments used to make the measurements were all proved to be flawless. So, we can take these numbers at face value and analyze what they imply one by one. In general, the LDL cholesterol level of men is higher than women and so they end up getting more heart-related diseases. Women doing better in language skills assessment is not surprising. The efficacy of drugs measurement showing men being less responsive could be explained by the bigger physical structure of men, which would require more medication to be effective. Women being able to leg press more weight than men is a little surprising. Perhaps these women living in Venus had developed a culture where they exercise more, which could explain their higher strength levels. When it comes to height, men being taller than women is quite common. So, the chief scientist who examines all these data finds them matching his expectation and consistent with common wisdom. So, gives his seal of approval and goes home. Are all these analyses correct?

The great granddaughter of Sherlock Holmes, who is a researcher in the field of biology, finds these conclusions disconcerting. When she starts thinking through them, she figures several valid factors were not considered while these analyses were conducted and so jumps into more sleuthery. When she digs through John's and Joy's families, their heredity, the way they grew up and so on, she is able to understand that many of these conclusions are incorrect. So, she publishes a critique pointing out how these differences were the results of factors nothing to do with gender. Let us investigate some of her arguments. First, she points out that these Mars and Venus residents were created using cloning techniques, which will eliminate the standard levels of variations you will find in the general population. Instead you will find the residents reflecting the original qualities of John and Joy in Mars and Venus residents respectively. Beyond that finding, we can classify her reasons into four different types.

Genetic Reasons: While John's family back in Finland were taller, had higher levels of LDL cholesterol, and capable of metabolizing drugs well, and Joy's family in Thailand were of opposite characteristics with correct levels of LDL, shorter in stature and with a body type that doesn't metabolize drugs well, then all the variations noticed in the data could be explained without any reference to gender differences.

Societal Reasons: Thailand's Joy might be very fluent in Thai and English. So, the Venus residents were well versed in both languages. But Finland's John was very fluent in Finnish but not in English. So, since the test was administered in English, the Martians ended up doing poorly on the test.

Environmental Reasons: Just like gravity that exists on the surface of earth, there are similar gravitational forces in existence in Mars & Venus as well. But the force is very weak in Mars, while in Venus it is much stronger than Mars, though not as strong as it is on Earth. So, it is possible for people living on Mars to grow taller over a century.

Combination of Environmental + Genetic Reasons: Since John's family members were originally taller, it might have compounded the effect on Mars allowing the men there to grow even taller. But the same lack of gravitational force results in their body weight imposing a lot less force on their legs. On the other hand, the shorter beings that are clones of shorter Joy, were also forced to deal with lot more weight on their legs due to the higher gravitational force that exists on Venus. So, their legs also grow much stronger over a century.

While Joy was traveling through space, she experienced a small dose of radiation that results in a gene mutation in her DNA. This mutation results in lower levels of drug metabolizing capability resulting in less drug efficacy on those clones compared to John's clones.

Thus, there could be a totally different set of reasons that explain the differences in the data collected on the two planets. But since the public as well as the experienced scientists hold pre-conceived notions of how things will be, the variations found in the reported data are all explained away by pigeonholing them into a standard set of understanding and explanation. What this thought experiment brings out are those

Whoopi Goldberg

confirmation biases.

Biases of so many kinds are present in abundance sprinkled across every race, language, skin color, gender, nation, and religion variations all over the world. Whoopi Goldberg is an American actress, born and brought up in the US, an EGOT winner, meaning she has won Emmy (TV award), Grammy (Music award), Oscar (Movie award) and Tony (Stage award) awards for shining bright in four different areas of performing arts, has done work in movies and TV and has even been a talk show host. Despite all these facts, in an interview she said that even now thinking that they are complimenting her, people talking to her will say, "Your English is really good!" Since she was born and brought up in the US, what is surprising in the fact that she is able to speak good English? The reason people make that comment is because of an innate but incorrect understanding they hold in their mind that since she is an African American woman, her language skills may not be that good. Most people carry good or bad opinions about other groups of people that are generalizations to help them understand and navigate the world they live in. But most of them don't always realize that they may be applying that generalized caricature opinions to a specific individual in front of them to whom many of those characterizations may not apply even though they belong to that group. The notion of dealing with each person as an individual and not as one of a group is hard to internalize in real life for most people.

Legal scholars have worried about innate biases working for or against a defendant in the justice system. It is a common practice all over the world for the defendant's attorney to ask the defendant to appear in the court with a respectable look and feel. This transformation may include dressing up the defendants in a suit and tie or some such business attire, making them speak in certain way and so on with the hope of influencing the judge and jury that will decide the fate of the defendant. While asking a defendant who is a gang member to behave respectfully in the courtroom may sound appropriate, we can't deny the fact that ideally the defendant should not be judged by how he or she behaves in the court during the trial but strictly based only on what happened when an alleged crime took place or the equivalent objective facts. But since the plaintiff, defendant, judge, jury, attorneys are all going to be in the courtroom together, how do you ensure that innate biases don't contaminate the proceedings? In the 1996 Hollywood movie A Time to Kill, during a tense courtroom drama scene, the defendant's lawyer making a passionate speech defending a black father who killed the white rapists of his 10-year-old daughter, tells the jury to close their eyes and listen to a story. He then narrates the attack conducted on the little girl in great detail and ends the narration with "Now imagine that she is white", to make the all-white jury feel the pain. This is a quick thought experiment that the jury is asked to perform, that helps the African American defendant father get acquitted. While this may not be feminism specific, it makes you think how much race, gender, how one looks, speaks, acts in the courtroom can help decide his/her fate in the end. To eliminate all such biases, there have been proposals for redesigning the courtroom completely. In such a bias free courtroom, the judge and jury will never see the plaintiff or the defendant in person. They won't see the attorneys either. All the statements that people involved will have to make will be scrubbed off for accent, emotion, intonation and will be read by a computer in a flat tone. Anything that needs to be demonstrated will be shown via computer animation using stick figures or some equivalent representation that is devoid of race or dress/jewelry type cues. It will be really interesting to see how many past convictions or acquittals will switch sides when the same trial is conducted in such a sterile environment. Even asking attorneys, judges and even jury to imagine such a court room in a thought experiment may make them feel uncomfortable since we gather so much of information rightly or wrongly by what we see and hear. So, initially it may feel incorrect to remove such cues involved personnel will find "helpful" in the proceedings. But thinking through it further may help us understand how wrong it is to use those helpful hints while we make our judgement calls in the end.

Biases lurk around in places you simply won't expect to see. A good example is what is called a Shirley Card. Way back when color film and photo processing came into effect, the technology and process was so new that the technicians in the photo labs needed a lot of training and help aids. Since the technology was mostly used to capture images of Caucasians, Kodak that pioneered the field calibrated the development process to fit the skin tones of Caucasian women. One of the tools they developed and delivered to the field techs is a color photograph of a young brunette with an ivory tone skin, a fluffy looking dress with red lipstick. When field techs developed negatives and made prints using a photo printer Kodak sold, they were trained to make sure the color of human skin in their prints matched what was in the sample printout sent to them. Since the model, who worked for Kodak and posed for these standard skin tone photos in the 1970's was named Shirley, the photo was called the Shirley Card. You can still find this picture online easily with a simple search. Though before it went bankrupt in the early 2000's Kodak did print out newer cards with multiracial women posing for them, the calibration standards set by the Shirley Card have seeped into even some digital photography standards and so expert photographers complain that they could never get the skin tones of brown or black people properly in photographic images as well as they do Caucasians.

Returning to gender biases in particular, it is also possible that such biases can end up benefiting women at times. For example, there is a general understanding that women in leadership positions will handle difficult encounters and challenges in better fashion than men without getting angry and thus lead their organizations in a much better fashion. There is also an understanding that women who are political leaders won't be as corrupt as their male counterparts and won't abuse their power. This has been statistically shown to be true as well. But those statistics apply to the whole population and should not be blindly applied to one individual female leader. Be it US or India, I am sure we can list women leaders who are corrupt, egoistic and abuse the power of their offices. Still we can certainly say via statistics that most of the gender biases work against women without even an iota of doubt.

I know a female physician of Indian origin practicing medicine in the U.S. She is very bright, has more than a decade of experience in treating a wide range of patients by age or demography. Despite her success she says, "The expectation of an average American patient is that their physician is a white, 6-foot-tall male who is slightly old. I don't check off any one of those four boxes. So, when I enter the room to treat the patient I am always forced to start from a point of disadvantage. In this situation, if a 6-foot-tall, white, slightly old male physician enters that room, the patient will be predisposed to listen to that physician respectfully and follow his orders. The male doctor doesn't even realize this advantage he has most of the time."

Take the time to think through how many such micro-biases lurk in your own life experiences. You may end up remembering the instances where such biases worked against you putting you at a disadvantage. But we hardly ever notice such biases we carry in ourselves and because of those biases how we mistreat others we encounter in life or put them at a disadvantage. For example, there is a general bias in most parts of the world that makes people think that anyone who speaks good English is very intelligent. This bias is highly prevalent in many Asian countries. In countries like India where higher education is mostly taught using English, this could be statistically true. So, it is not surprising if a person who speaks English well turns out to be very intelligent. But that rule need not apply to every individual we encounter who speaks English well. I have personally interacted with dumb individuals who speak splendid English and brilliant individuals whose English was subpar. Examples from all over the world come to mind.

Even amongst people who speak the same language, researchers have found that women who speak in a higher pitch are not respected or taken seriously compared to their male counterparts. I came to know about the experience of a female attorney named Monica Hana practicing law in New York. Monica who is just five feet tall tends to look like a teenager in person. She is a fully qualified attorney who has argued and won several cases in her career. But since her voice tends to be on a higher pitch, starting from her attorney colleagues to her own clients, she noticed that many are not giving her the respect she deserves. She didn't want to be permanently at a disadvantage. So, she ended up consulting a voice coach and based on the guidance she received incorporated several subtle changes in her intonation and pitch. This approach resulted in a dramatic improvement in her career success. You can listen to her story

Malala Yousafzai

on the NPR.org radio podcast titled *Can changing how you sound help you find your voice.* The kinds of gender biases women have to overcome in the workplace is truly appalling.

You might have heard the news that Malala Yousafzai, originally from Pakistan was awarded the Nobel Peace Prize in 2011 making her the youngest recipient at the tender age of 17. Of course, we need to admire her courage and her continuing efforts to promote female literacy all over the world. But even in the twenty-first century, so many having to fight for girls' right to go to school to get an education, Malala getting shot in the face for saying this and later ending up with a Nobel prize are all clear evidences to the fact that the world has a long way to go to overcome its gender biases. It is indeed such a sorry state of affairs. Before this century ends, hopefully our grandchildren will wonder as to what revolutionary thing this Malala promoted to deserve such a high honor. Hopefully the time when this turns out to be a headscratcher will show up before our time ends in this planet or Mars or Venus.

--------- o0o ----------

Physical Universe

What we observe as material bodies and forces are nothing but shapes and vibrations in the structure of space.

– Erwin Schrodinger

Schrödinger's Cat

A book on thought experiments cannot be completed without this cat crossing our path. So, let us say hello to this kitten that has appeared in so many research papers, write ups and thought experiments for several decades.

In our day to day life, we can precisely say that at this instant John weighing 75kg is traveling from Allentown to New York in a car at 65mph speed whistling Dixie. But instead of John, if we dive down to atomic particles level, we are not able to make similar statements about the movement and direction of electrons and protons. Each subatomic particle seems to exist like a vaguely defined cloud. As early as 1930s a school of physicists studying these particles figured Newtonian physics that works for human being size objects is not adequate and developed quantum physics as a better alternative. As this field started gaining momentum then, the idea of wave functions that explain the subatomic particle characteristics was introduced. Using these so-called wave functions, they were able to describe the subatomic particles' location, energy level, velocity and other such characteristics as probabilistic functions. A school of thought named Copenhagen Interpretation was becoming popular in those days. That interpretation said when no one is observing, these particles existed and traveled in different directions simultaneously in varying probabilities. They also claimed that when a measuring instrument started looking at any given particle asking, "Hey dude, where are you and what are you doing?", this wave function instantaneously collapsed allowing the particle to appear as a clean entity precisely present at a specific location with a set energy level.

If we start describing John using such terms, we may end up saying he is traveling between multiple cities in the Northeast US simultaneously at different elevations as well as weights in various percentage probabilities. In other words, this will mean he exists like a thin cloud spanning two three states in the US. If you add up all those probability percentages, it will add up to 100. If you think this description sounds absurd and so turn your head to see this spanning John cloud, he will instantaneously hide his cloud avatar, switch back to his normal human form and appear to be traveling in his car from Allentown to New York City whistling Dixie again! Though it is impossible for human beings to behave this way in real life, subatomic particles do play peekaboo with us exactly in this fashion. Many have opined that this notion of something being present here, there and everywhere simultaneously seems to oddly resemble such ideas proposed by Hindu religious texts for mythical Gods. As per the Copenhagen Interpretation, the mere act of interacting with the subatomic particle, even to the mildest extent of just an instrument passively observing it to quantify its characteristics seems to alter its characteristics. Einstein wrote about this phenomenon along with two other scientists. Erwin Schrödinger who was one of their contemporary and a brilliant physicist, was discussing this idea with Einstein himself. Since Schrödinger thought this is all nonsense, he proposed a thought experiment in 1935 to convey his thoughts as to why this interpretation doesn't make sense to him.

Erwin Schrödinger

In this experiment an iron box that contains a live cat is described. Within the same box there is a glass vial containing enough quantity of a poison that could quickly kill the cat. In another little container located within the same box there is sample of a radioactive element. It could even be just one atom of that radioactive element. As per the setup, there is a 50% probability that the element will undergo radioactivity and emit a particle. There is a Geiger counter positioned inside the box to sense this radioactivity as soon as it occurs. Once it occurs the sensor will trigger a hammer that will break the little glass container holding the poison. Once the glass container is broken the poison will emit a noxious gas instantly killing the cat. It is safe to presume that the radioactive element, Geiger counter, hammer, poison are all beyond the cat's reach and so cat cannot accidentally trigger this sequence committing suicide. Only way cat will get killed is when the element puts out radioactivity. Till then the cat will remain perfectly safe.

In this experiment if we leave the box alone for one full hour, is the cat inside alive or dead? Since the radioactivity can take place anytime during the hour at 50% probability, we can't really say for sure what the state of the cat is at any given instant. If we want to know the truth, we can of course open the box and see inside, and we will know immediately if the cat is alive or dead. But until we open the box, going by the Copenhagen Interpretation, we need to concede that the cat is simultaneously alive and dead. We can all come to an agreement that this is not possible and so the whole interpretation doesn't make sense is Schrödinger's argument.

Starting from Einstein in that era to present day comedians, this thought experiment has attracted so many inspiring them to discuss it in various forums. So, even today we can often find this cat making appearances in jokes, articles, TV programs and other discussions.

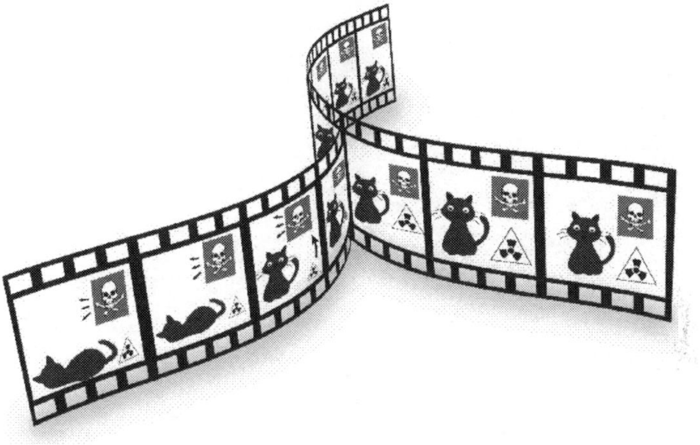

We can run through some of the interesting responses this thought experiment had evoked over the years. After the 1930s, the notion of multiverse was introduced. It argued that instead of one universe, there could be multiple universes in existence simultaneously. Proponents of the multiverse notion suggested that presuming that the cat is alive and dead at the same time is not fundamentally incorrect, when we open the box the universe we are in, splits into two. In one universe the cat is still alive while in the other universe it is dead. People living in one universe can never interact with or see what is going on in the alternate universe. So, people living in each universe truly believe what they see is the only reality. Thus, this leads to self-consistent worlds that are not colliding with each other is their explanation.

There is a different explanation that goes like this. Schrödinger has cleverly mapped the state of a subatomic particle and its characteristics to something that is as big as a cat or John traveling between cities in the Newtonian world we understand. But even before the human beings interested in finding out whether the cat is dead or alive open the box, the cat in this scenario itself is serving as an observer. Even before that the Geiger counter sensing whether there is any radioactivity inside the box is another observer in the story that is even closer to the atomic level activity that may be taking place. Since that counter is observing the element, right then and there the wave function we talked about earlier collapses resulting in the quantum phenomenon vanishing from the picture leaving us with an atom in more of a Newtonian state. So, experts in this area argue that it is wrong to think of this whole box and the human being outside together as one experimental setup.

Alternately you can think of the wave function collapsing once when the Geiger counter senses radioactivity, another time when the cat dies/remains alive, yet another time when the observer opens the box and so on. You can also view it from yet another angle saying if you run this experiment one thousand times, five hundred times the cat will be alive and the remaining 500 times it will die. This is all the radioactivity indirectly shows you through this experiment. What do you think?

If you step back and look at it from another perspective, we can even claim that as long as we are discussing this thought experiment, this Schrödinger's cat will remain alive for several more centuries.

Survivor vs. Survivor

I am often amused and puzzled by the gaping differences that tend to exist between reality and perception. A while back a good friend of mine had to go through treatment for subdural hematoma, which is blood leakage outside the brain and so is a serious condition that must be treated quickly. Not a pleasant experience to go through to say the least. A fix for the issue is guiding a wire from the groin area to the patient's head (with the help of a live scanner image showing where the tip of the wire is) and then plugging the holes that may be causing the leak. The surgeon who does this procedure apparently describes himself as a *Navy Seal* meaning one who has to be fantastic in his job and can't fail. While physicians are often (but certainly not always) known to have such a *God Complex*, our society tends to give them a pass since they save human lives.

If I start telling the general public about what an engineer in our company does to get a CPRI (Common Public Radio Interface) to work, I will lose them in less than a minute. While a lot of work (but certainly not everything) in the semiconductor industry is extremely inventive and complex, since our work gets buried several layers underneath the interface of an Android phone that the general public interacts with, not many people outside the industry understand what goes on under the hood and our engineers hardly ever get the reverence the subdural hematoma fixer gets in society.

If you don’t like this norm, how will you go about designing a system that evaluates anyone's contribution more objectively? Say surgeon vs. semiconductor industry engineer vs. carpenter vs. teacher? How can society's perception be objective?

Suppose you ignore the automatic value added to "saving human lives" aspect of the subdural hematoma procedure and look at what actually occurs. This surgeon didn't invent the procedure and so should be given no points for creativity. At best he is a technician who is executing a procedure he has been taught. Is it any different from, say, a machinist making a metal widget or a carpenter making a wooden chair following a predefined procedure, using the tools given to him/her?

Note that people working within one system all their lives will often tend to see their own field as fantastic or extremely complex. I have many friends who are physicians. I won't be surprised if they wonder if I have enough understanding of the value they bring to society to write this section. I am sure many of my friends in the semiconductor industry will probably give a wry smile and say our work can never be understood in full by the general public.

This disconnect was hilariously exploited in a comical scene. Curb Your Enthusiasm is an American TV comedy show developed by Larry David who also co-created the Seinfeld sitcom that ruled the 1990's. In one CYE episode, an old Jewish Holocaust survivor meets a young participant of the CBS TV channel's reality TV show Survivor in a dinner party. Mayhem of the hilarious kind ensues as seen on this clip:

https://www.youtube.com/watch?v=In2XfN3hIi4

As per many books written and research done on human perception, what we experience and see up close will have a much bigger impact on our psyche than a lot of bigger/more important things going on further away from us. So, it is no wonder that the CBS show's survivor in this clip truly believes that he has gone through a lot of hardship (no access to gym, snacks or Wi-Fi and having to wear flip-flops) though he is talking to a Holocaust survivor.

I am certainly not envious of physicians since the reverence from the society they get comes with its own negative consequences. They get sued if a procedure goes wrong though it may not be strictly under their control. A semiconductor industry engineer getting sued for something that went wrong is extremely rare. The B2C (Business to Consumers) interface where physicians deal with a large number of the not-so-well-informed general public, that sometimes gives them the God Complex, can also result in burnout since the doctor-patient communication is so difficult. On the other hand, my industry being B2B (Business to Business), in our professional lives, we deal with a much smaller number of people repeatedly who all tend to be well-educated experts in our field. As a result, communicating our ideas becomes a lot easier. If there are long discussions, it is because the subject under discussion is complex and not because the other side is not understanding what is being said. So, I certainly don't envy physicians and I am happy where I am. But the disparity in how society views the complexity involved in what we do is phenomenal.

There are similar disparities in perception even on the flip side where society looks down on something. Take the recent Paul Manafort sentencing in the U.S. (He is a convicted felon who was President Trump's campaign manager.) For some 55 million dollars' worth of tax evasion, money laundering, colluding with foreign entities, etc. he received less than 4 years of prison sentence. Yet there are hundreds of people thrown behind bars for decades for selling a few grams of drugs on the street or for stealing $100 in a robbery. Thus, white collar criminals are given a pass while an armed robbery (even when no one was physically injured) or selling drugs is punished an order of magnitude more harshly by countries all over the world. How would you design a system that objectively evaluates the harm done by different actions? What should make a difference in how to evaluate a white-collar criminal vs. how to evaluate a street criminal? How can we be sure that we are observing/evaluating what we see or experience consistently? This question was answered by a thought experiment proposed by physicist Eugene Wigner, a Nobel laureate, in the early 1960's.

This TE called Wigner's Friend envisions Wigner and his friend conducting experiments on a single photon. This particular photon happens to be a polarized one. As per quantum mechanics, until the instant the measurement of the polarization takes place, the same photon can be in horizontally and vertically polarized states simultaneously in a state called superposition (recall the Copenhagen Interpretation). As soon as the measurement process sets in motion, the photon will switch to either horizontal or vertical polarization. In the TE, Wigner's friend conducts the measurement far away from Wigner, who is simply watching his friend. Since Wigner is far away, he is not part of his friend's measurement process and so will presume the photon is in superposition (i.e. in both horizontal and vertical polarization status) while his friend will see it clearly in one of two positions and record it so at the same time. To make it more interesting, we can even say Wigner is simultaneously conducting another experiment to prove the superposition status of the photon. We know that as long as Wigner does not try to measure the polarization but proves it in another way, he will succeed in showing the same photon to be in superposition, while at the same time his friend is measuring and recording how the photon is polarized.

This thought experiment shows that there is no objective reality that would be agreed upon everyone involved in a situation or a phenomenon. In fact, very recently, Caslav Brukner, at the University of Vienna in Austria, managed to run an actual lab experiment where six entangled photons were used to create two conflicting realities. In one part of the experiment, the polarization was measured, while simultaneously, another part of the experiment proved superposition. Showing that these two irreconcilable realities can coexist, Wigner's idea is demonstrated in the lab, while posing challenging problems to future physicists.

This concept has been explored in depth by the famed Japanese movie director Akira Kurosawa in his classic named Rashomon. Multiple characters in the movie witness or participate in the same event, but each describes it very differently. We should note that in the case of Survivor vs. Survivor, we are talking about two totally different experiences being perceived as equal by two different observers whose perceptions were shaped by their past experiences. In Wigner's Friend TE and the movie Rashomon, the situation is simplified further where the observed phenomenon is the same. Still two or more observers of the phenomenon can't even agree on what actually occurred. Since it is not only movie fodder, but an idea for a thought experiment that has been proven by physicists in the lab, perhaps it is impossible to develop a system that can observe and assess situations objectively to provide judgments that everyone will agree on. Would you be able to create a 'normalization factor' that can be applied to different experiences so that they can be evaluated on a consistent, uniform scale?

Riding a Light Wave

Einstein was a sixteen year old boy in 1896. At that age he had wondered how it will feel if he could sit on top of a beam of light and travel along with the beam at the speed of light. A decade later this thought experiment had led him to the Theory of Relativity, still considered a magnum opus of the human mind. In his autobiography he has described as to how he used to wonder about this theoretical experience. He says the initial grain of this thought was planted in him when he was reading a story by the great Sci-Fi writer Aaron Bernstein, who was a nineteenth century German author and scientist. In his story Bernstein had talked about traveling inside a telegraph wire at the speed of electrical signal. He also wrote in his stories about a faster than the speed of light traveler. We know the Sun is so far away from earth that the light emanating from the sun takes eight minutes to reach us. So, when we look at the sun, we are not seeing how it is at this instant but an image of the sun as it existed eight minutes before. Similarly, when we are looking at a star that is one hundred light years away from us, we are actually seeing how that star appeared one hundred years back. So, if we can somehow instantly travel one hundred light years away from earth and then look back at earth from there, we will see a hundred year old earth. First World War might be just starting up and Second World War may be years away. If we can similarly jump to a point that is seventy million light years away and look back at earth, we will see a planet inhabited by dinosaurs. Thus, to see any past event taking place on earth, all you need to do is jump to a point in space that is so many light years away so that you will be able to look back at earth and see what was going on at that time in history since you will be catching the light still carrying those images and traveling through space. While this may not be exactly time travel as described in popular fiction, these are the first ideas of time travel (or looking back at the past) that were scientifically accurate. In addition to these stories, Einstein also read the

Albert Einstein

works of David Hume, a Scottish philosopher who proposed that all the scientific principles and laws that we think exist in the universe and are being discovered by scientists are actually being originated within the human mind. This allowed or prompted Einstein to create his own universe and constructs in his own mind that may not gel with existing (read Newtonian) laws and understanding. Let's see how his thought process worked with respect to the speed of light.

Let us say you stopped at a gas station to fill up gasoline on your way home from work. As you are about to start pumping, you see your friend Jane, who had just finished filling up, driving away from the adjacent pump. If you want to catch her to say hello, all you need to drive in the same direction as her at a faster speed. Depending upon how fast you are compared to her, you will catch up with her sooner or later. The gas station attendant seeing you driving away from him fast, might even see you catching up with her within his field of vision. This is all just routine stuff and completely normal and possible. Correct?

Now let us say the person who drove away first is not your friend Jane but Prof. Einstein himself and he is driving his car like a maniac at the speed of light, which is equal to 300,000 km/s. You are determined to catch up and say hello Einstein today. So, you are hitting your gas pedal on your car that is also designed to run at that high a speed. Hopefully the roads are all empty to allow you to drive like both of you are on road rage. As you accelerate, you reach a top speed of 290,000 km/s you remember that you pulled into the gas station to fill up but started driving to catch up with Einstein before you pumped gas. Since you are low on gas and don't want to be left stranded somewhere in space where there is no AAA service to bring you gasoline, you give up, turnaround and go home. Next day when you return to the pump (remember you are still low on gas and so need to fill up), the gas station attendant who recognizes you from yesterday starts chatting as you are filling up. He says, "I saw you driving away really fast yesterday to catch up with Prof. Einstein. Since you were also accelerating close to the speed of light, you must have been very close to catching up with him. Correct? If only you had enough gas, you would have accelerated enough and caught up with him within another few million miles." When you hear this remark, it will sound quite weird to you. When all this chasing occurred yesterday, since you sped up at a rate of 290,000 km/s, the stationary attendant would have seen Einstein speeding away only at a mere additional 10,000 km/s compared to you. But to you driving at 290,000 km/s, Einstein's car ahead of you will still be appearing like it is moving away at 300,000 km/s speed from your vantage point. So, while the attendant might have felt that you were pretty close to catching up with Einstein, a totally puzzled you would have responded with, "What are you talking about? I had no chance whatsoever. Never knew Einstein was such a maniac."

This total disconnect between what you experienced and what the gas station attendant saw is explained by the weirdness imposed by the speed of light on this universe. We are certainly not used to it in our day to day life. As per Maxwell's equations, the speed of light is a constant anywhere and everywhere. (To be precise, we should say, speed of light is a constant, when we observe without undergoing any type of acceleration.) So, whether you are standing on the ground like the gas station attendant or driving at 290,000 km/s speed chasing Einstein or traveling in a plane or a rocket towards or away from the tip of a light beam shining into space, the speed with which light is traveling is not going to change. Since we don't experience this phenomenon consciously in our day to day life at the small speeds in which we operate, all this may sound quite weird. But just like the notion of infinity we encountered in Hotel Hilbert, speed of light is also a strange concept. If Einstein's car racing ahead of you really reaches the speed of light, he will also encounter several other really strange experiences. First his weight will keep increasing rapidly as he approaches the speed of light. At the same time his body will continue to shrink into a very small point. If he somehow manages to accelerate the car, even then finally reaching the actual speed of light, while his weight would have become infinity, time will come to a standstill. This is the reason because of which no one can travel faster than the speed of light anywhere in the universe.

Einstein was living in the city of Bern in Switzerland during his patent office stint. Legend is that he could complete his tasks at the office so easily and quickly that he had ample time available to work on his physics problems. He was returning home one evening riding a tram in the same street seen in the picture above. As he was riding away from the clock tower, his mind went back to his thought experiment once again. So, he wondered as to what will happen if this tram can move at the speed of light. Suddenly several things became clear to him. If the tram is indeed moving at the speed of light away from the clock tower, the visual image of the clock tower that shows the passage of the time as he moves away can no longer reach his eyes. That image is also traveling from the clock to his eyes only at the speed of light. Right? Since he is moving away at the same speed, new images can no longer reach his eyes. So, to him it will appear as if the clock tower clock has stopped advancing. But a wristwatch he is wearing, which is moving with him, will show seconds ticking away normally as it always did. Starting from this understanding clicking into place, Einstein did several mental physics and mathematics somersaults, to form a new well-lit universe that had different clocks running at different speeds in different locations never seen in the history before. This sublime and fantastic explanation named Special Theory of Relativity threw the Newtonian world where clocks always ran at the same speed everywhere maintaining a linear flow of time, unceremoniously into a pile of discarded trash. Recently when I was in Switzerland, I paid a visit to that clock tower in Bern to pay my homage. Still stands tall, well-maintained and beautiful looking, quietly knocking away seconds marking time as long as you promise not to run away from it at the speed of light. A small house where Einstein used to live is being maintained as a museum now. Walking around the museum felt like a religious pilgrimage.

In our daily lives we move like snails in a speed range too far removed from the speed of light. So, we never encounter such crazy phenomenon of clock stoppage and time dilation. But when we are interacting with satellites, due to the extraordinary speeds with which they are flying compared to us, effectively fixed in a geostationary orbit, this time dilation effect slowly starts to show its head. For example, if the GPS equipment you use in your car has to work normally, the implementation must account for this time dilation. Otherwise clocks running on the earth compared to the ones running on GPS satellites will introduce about 37 microseconds difference each day. This time drift alone can render all the GPS implementations, be it on your phone or as a standalone equipment mounted in your car or incorporated as part of a plane's avionics, useless within a few hours. Immediate effect will be the Main Street in your town being shown as located few miles away from where it actually is. This is enough to cause massive confusion or to render GPS unreliable.

If you take the mass of an object and multiply it by the square of the speed of light, the resulting number gives you the amount of energy stored in that object. The world-famous $E=mc^2$ equation expressing this notion is one of the elegant results of this thought experiment. This simple equation tells us how large an amount of energy is stored in the universe. Take another look at that deceptively humble looking equation. How can someone tie energy, mass of an object and the speed of light, that all appear as totally disconnected ideas into one equation? No wonder the world still considers this creation a monstrous achievement even after so many decades since it was presented to the world.

Floating on an Elevator

Few years after publishing his Special Theory of Relativity, Einstein asked us to dive into another thought experiment. Imagine very tall sky scrapper. You are in an elevator inside that building traveling at a very high speed from its 1000th floor down to the ground floor. If that elevator descends really fast, you may experience the gravitational force a little less and may float in the air briefly. When it finally decelerates as it approaches the ground to come to a full stop, you will stop floating and will get your feet back on the floor of the elevator feeling the full force of gravity once again. You might have heard about Hollywood movie makers using this technique to shoot scenes of weightlessness in space action movies, such as Apollo-13. They build the whole set to shoot the scene inside a large aircraft. The aircraft is then flown several thousand feet up into the sky with the actors and film crew positioned inside the set. Once they reach the maximum height possible, the pilot does a nosedive flying the plane very quickly down towards the earth. As the plane rapidly descends towards the earth, the entire set inside with the film crew and actors experience weightlessness for few seconds. It is never more than a minute or two as that is all the altitude that the plane can safely operate in while descending rapidly. Once it is close to earth, the pilot stabilizes the plane and lands it or takes it up for another loop. All the scenes that the movie makers want to show that is apparently happening in the weightlessness of space is recorded during those few seconds or minutes of the repeated quick descents. The elevator Einstein describes in his thought experiment introduces the same effect of weightlessness.

As a next step in the experiment, instead of imagining yourself inside this elevator in a skyscraper, imagine you are floating inside a spaceship beyond earth's gravity. Since there is no gravity to force us to keep our feet on the floor of the spaceship, our default position will be floating in the space inside the spaceship. Now if this spaceship is fired up and accelerates towards its destination somewhere out there, your floating state will come to a quick end and you will be forced to plant your feet on the floor or walls of the spaceship, depending upon in which direction it is accelerating. You will experience a force very similar to gravity that will keep your feet properly planted on the spaceship. You may feel your weight also as you lift your hand or turn your body around. When the acceleration slows down, even though the spaceship is still traveling at a very high constant velocity, the gravitational force you experienced will vanish and you will start floating inside the spaceship again as before. But as and when the spaceship accelerates, you will experience gravitational force proportional to the rate of acceleration. Thus, the equivalent of the gravitational force that you experience is nothing but you perceiving the force of acceleration as you travel in space.

This spectacular understanding Einstein gave us through his thought experiments could be pushed further. Visualize what we describe as space-time as a piece of stretched fabric. If you throw a ball into this stretched fabric, it will depress a bit. In a very similar fashion, when a large mass as big as a planet is located on the fabric of space-time, it tends to deform and depress a bit. Newton's first law of motion says that in the empty space of our universe, a body that is moving in one direction will continue to move in the same direction forever until disturbed by some external force. Since there is nothing in space, they are not going to slow down due to wind drag or any such force. If such masses go close to the depression shown in the picture, its path will change from a straight-line to a circle. The size of the depression created in space-time is governed by the mass of the object in the center. Bigger the mass, larger the depression. In the little corner of the universe where we happen to live, the large mass distorting the space-time is the sun. All the planets including the earth are stuck in the space-time distortion the sun created and so continue to travel around it in planetary orbits. The reason moon moves around the earth is due to the same reason. Einstein's this elegant explanation answered a lot of questions like why do planets go around the sun, where do they get the energy to keep circling the sun endlessly and why don't they peel off and move in different directions.

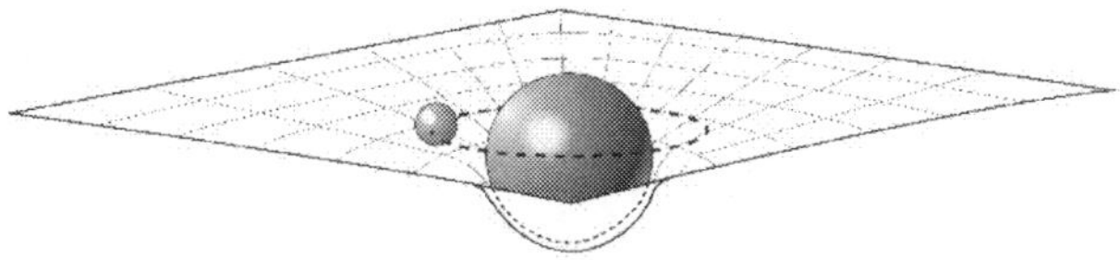

These thought experiments and the explanations Einstein provided as to how our entire universe works are all the heights of achievements reached by the human mind. Almost three decades back, a brother of mine who is a little older than myself in age but a lot older than I am in maturity, pointed all these details to me during our friendly discussions on these topics. I never really understood the nuances well during my school days. Perhaps my focus was more in learning the material and writing the tests to move on to the next class than to truly understand and admire how elegant these ideas are. But when I was shown the beauty of these concepts, I felt so thrilled, humbled and honored at the same time. I still fondly recall the buzz I felt for days. I am not sure even if addictive drugs can give you such a sustained buzz and pleasure for such a long time. I truly believe that it is our duty to show and explain these miracles to the generations that follow us, so that they can derive as much pleasure understanding these concepts as the previous few generations did.

--------- o0o ---------

Thinking About Commerce

The lessons taught in great books are misleading. The commerce in life is rarely so simple and never so just.

– Anita Brookner

Four university students skip their classes one day and head out to watch a new movie. It so happens that the professor teaching the class they skipped unexpectedly conducts a pop-quiz on that day. These four come to know about this important test only when they return that evening and hear about it from classmates. They don't feel like accepting responsibility for cutting class and so come up with a scheme to cheat the professor. As per their plan they approach the professor next day and say that they headed out to study in a better library located in the city and were returning to the class when the car had a flat that prevented them from returning on time. On hearing this story, the professor says in a very understanding tone, "Sure, since all four of you have promptly told me the truth, let me give you another opportunity with a makeup test." The students feel glad that they managed to fool the professor. As they also managed to study a bit overnight, they feel better prepared than the other students who took a surprise pop-quiz the previous day. Professor makes them sit on four different rooms and gives them the question paper. It has questions related to the lesson worth just 5 points. The remaining 95 points are for one simple question: Which one of the four tires went flat yesterday?

This joke is one manifestation of the famous Prisoner's Dilemma thought experiment.

Prisoner's Dilemma

Circa 1950, Merrill Flood and Melvin Dresher, two scientists working for Rand Corporation proposed this notion. Subsequently Albert Tucker, a mathematician from the Princeton University formalized the idea, named it Prisoner's Dilemma and made it part of the game theory oeuvre. The underlying idea demonstrates that people who are very rational and know that cooperatively working together with your counterpart will benefit both, will not cooperate often. This is

Albert Tucker

despite knowing well that not cooperating will be to their disadvantage.

In this story there are two prisoners. Let's call them Alan and Bryan. Together they have robbed a bank at gun point. Though the police have caught them, police don't have enough evidence to convict them. So, they start interrogating both simultaneously but separately in two different rooms. Though the interrogation continues for several days, Alan and Bryan never get to talk to each other privately. Both are clever, educated, rational beings. But since they didn't get to sync up on their story and strategy on how to handle the police interrogation, and are also not allowed to exchange notes on how the interrogation is proceeding, they are a little nervous. Each wonder as to what the other person is going to say. So, the detective interrogating Alan says, "Alan, we know very well that you were at the scene of crime. So, we are going to lock you up at least for a year in a maximum-security prison. But if you confess, become an informer and help support Bryan's conviction, we are willing to let you go free. Then we will put that Bryan behind bars for ten years. If you decide to keep quiet and Bryan turns against you, you can be sure we will turn you into dead meat. You are going to spend the next ten years behind bars. What do you want to do? You better confess and help us put that Bryan guy in prison."

After thinking through this offer, Alan asks, "Ok. What if Bryan also decides to rat me out?" The detective says, "Then we are going to lock both of you up for five years each for sure."

Another detective talking to Bryan makes the same offer to him as well.

If we rationally think through the ways in which this story can end, we will quickly discover four possibilities. We can tabulate them as shown below.

		Bryan	
		Keeps Quiet	Rats Out
Alan	Keeps Quiet	A : 1 Year B : 1 Year	A : Freedom B : 10 Years
	Rats Out	A : 10 Years B : Freedom	A : 5 Years B : 5 Years

The top left square indicates the scenario where both Alan and Bryan decide to keep quiet. Since police will not have any evidence or confession to lock anyone up for a longer period, each one will stay in prison for a period of one year, which is the minimum time for which police could lock them up with the minimal evidence they have. If both decide to talk, trying to rat the other person out in exchange for a lenient sentence for themselves, we end up in the lower right square. In this case each get 5 years of imprisonment. The other two squares (top right and bottom left) indicate scenarios where one person rats the other out, while the other person decides to remain quiet. So, the informant gets away free while the other prisoner gets ten years.

If you are one of the two prisoners, will you keep your mouth shut, hoping the other prisoner will do the same and both of you will get away with a one-year sentence? Or will you rat the other person out in the hope that you can get your ticket to freedom? As a rational academic analyzing the problem with these four different possibilities, one can easily conclude that the best option for the prisoners is to stick to the top left square. Both get one-year prison term. Combined together this is a total of 2 years of imprisonment. This is the least of the combined prison term amongst the four squares. All the other three squares show situations where the combined total of prison term is ten years. Since no one wants to be in prison, this is clearly the optimal solution from the prisoners' point of view.

Next let's analyze the same possibilities from the individual prisoner's point of view. Let us say you are Alan. Bryan in the next room may keep his mouth shut or sing like a cannery to the detectives pressuring him and rat you out. If he is going to keep quiet, you will be better off providing evidence to get him convicted so that you can get away Scott free.

What if Bryan rats you out? Even then, you becoming an informer is the better choice. Otherwise you will be the sucker stuck in prison for ten years. Thus, though it appears obvious for an external analyst that both prisoners should keep quiet to minimize the prison term, when you look at the same situation from the individual prisoner's point of view, betraying the partner seems to be the correct strategy. This paradox is what this thought experiment is trying to illustrate.

As Alan, you may think that betraying your partner this way may land you in a lot of trouble later. Bryan may get out of prison one day and come after you for revenge. Perhaps even Bryan's friends or family outside the prison may come after you when they learn as to what you did to get out. Even if you are not worried about attacks and revenge, you may think your own ethics and morals won't allow you to betray your partner even though you both committed a crime together. This issue has been analyzed for decades by experts in the field of Game Theory. To keep the initial analysis simple and objective, those moral quandaries are set aside to get to the gist of the issue. When you add such parameters and allow them to influence the analysis, the characterization of the problem will change. In the second level analysis, academics have looked into them as well. But let us look into the first level analysis excluding such factors some more.

Though this is posed as dilemma between two prisoners trying to get their prison sentence reduced, there is a lot of connection between this thought experiment and the world of commerce. Starting from two young girls selling Girl Scout cookies in your neighborhood to multinational companies competing with each other all over the world in billion-dollar transactions can all get into this dilemma. We can consider MNCs (Multi-National Companies) Honda and Toyota selling cars all over the world as the equivalents of Alan and Bryan respectively. If both companies decide that they are not going to change the price of their car offerings and are also not going to rise the marketing budget for next year, they both will end up with similar revenue/profit as they had in the previous year. This scenario will map to the top left square where both prisoners decide to keep quiet.

If Honda decides to get aggressive and start advertising big time while giving discounts to improve their sales, it will increase their sales while adversely affecting Toyota's sales. As long as Toyota also doesn't start implementing the same techniques to promote their sales, situation maps to lower left square. Alternately, if Toyota gets into a massive advertising campaign and discounts to promote their sales while Honda continues its previous year strategy, we end up in the top right square. If both Honda and Toyota get into more advertising and discounts simultaneously next year, this benefits the consumer (who is equivalent to the detective) and we will end up in the lower right square.

Thus, this thought experiment can be easily mapped to advertising budgets, discounted sales, introducing a new technology and many other such variations introduced in the marketplace to gain advantage in commerce. It is not even limited to different vendors competing with each other. Let us say you are picking up a piece of pie in the flea market in your town. If you simply pay whatever the seller asks you to pay without trying to bargain, you get the pie without having to put in a lot of effort but at a possibly higher price. Since both you and the vendor got along to get the sale completed quickly, this situation maps to the top left square. If you bargain endlessly to get the best deal possible tiring both yourself and the vendor out before you make the purchase, you land in the lower right square. Alternately, if you do bargain a bit but not a lot, you will end up in one of the other two squares.

A British TV show named "The Golden Balls" thrusts this dilemma right in the faces of the two competitors participating in the show. While the two participants represent the two prisoners in the thought experiment, the host of the game show becomes the detective negotiating with the prisoners. In the initial rounds the host asks various questions, participants answer and the prize money they both win gets consolidated into one box. The Prisoner's Dilemma notion is brought in only at the end when the participants must decide how they want to share the winnings. Let us say they have collected a sum of $100,000 by the time they reach the last round. At this point both contestants will get two gold colored balls each that look identical from the outside. If you open the ball, you will see "Split" written in one, while the other has "Steal" written inside. Both contestants need to choose one ball out of the two given to them. Effectively they are making a choice to either split the winnings with the opponent or to steal all the money for themselves. Contestants won't know what choice the other person made. Once they make their choice, they cannot go back. When the host tells them to reveal their choices, they both need to open the ball they had chosen and show their choice to the other contestant and the viewers simultaneously.

Let us say the characters Alan & Bryan we encountered before play the last round of this game. Depending upon which ball they choose, the game can end in four different ways, just like the PD case. If they both choose "Split" balls, it indicates that they are cooperating with each other. So, the prize money will be split into two halves with each one getting one half.

If they both choose "Steal", neither one gets any money. Host takes away the entire prize money leaving both contestants with nothing in the end.

Alan \ Bryan	Split	Steal
Split	A : $50,000 B : $50,000	A : $0 B : $100,000
Steal	A : $100,000 B : $0	A : $0 B : $0

If one contestant plays nice and chooses "Split" while the other contestant plays the devil and chooses "Steal", the person who played nice will go home empty handed, while the one who chose "Steal" will get the entire prize money becoming the winner of the show. Following figure shows these four combinations.

This show was on the air for several years. After the show is over, no one is going to beat up the winner that chose "Steal" and took away all the money. The contestant may never ever meet for the rest of their lives. In other words, they never need to interact again and so can treat this as a one-time interaction with the other contestant. Since both contestants understand these facts well, they will be pushed to play to go home with all the prize money. Thus, this scenario maps to the Prisoner's Dilemma problem exactly. Since they are sitting right in front of each other during the course of the show, they are allowed to talk to their opponent before they make the final selection of the ball.

Often in this show, we will see a friendly, non-threatening lady on one side. She will claim that she is a God fearing, very ethical person. She will talk about her educational qualifications, says that she comes from a respectable family and is a mother of two young kids, and so on. The opponent may be a large gruff male. So, she will plead with the tough looking dude saying, "Just looking at me, can't you see I am a God fearing, respectable person? I promise on my children's head that I will chose "Split". I beg you to choose the same so that we can split the money. This will allow me to take my children on their dream vacation that I can never afford otherwise." The scary male might melt at this pleading and wanting to show how we all can get along together to form a happy world, he will pick the "Split" ball. Now when the game show host finally asks them to reveal their choice, the lady will show that she chose "Steal", gather all the money and walk away the winner, leaving the tough guy to hang his head in shame for having believed that lady.

If you play this game, will you trust your opponent? Or will you try all you can to somehow make your opponent believe that you are going to choose "Split" and then sneakily choose "Steal" to cheat that person out of all the money in the pot? Think about this choice you will make carefully, write it down and then view the YouTube link provided below to see what a clever contestant did in one episode of the show.

https://www.youtube.com/watch?v=S0qjK3TWZE8

In the next chapter, we will look into scenarios where the interaction between two participants is not "one and done" but tends to be iterative. It will be quite interesting to see how people's behavior changes when they know they need to interact with the same participant repeatedly. This is also common in the world where participants routinely need to trade with each other. What tricks and strategies the two participants will use in that situation, what is the role of good human qualities we tend to value such as ethics and loyalty in that case will all be quite interesting to analyze. These are not just imaginary cops and robber stories. As we will see next, an economist has even received a Nobel Prize digging down deep into this problem.

--------- o0o ----------

Effects on Economics

The curious task of economics is to demonstrate to men how little they really know about what they imagine they can design.

– Friedrich August von Hayek

Years back when I was a school age kid, I was traveling with my family by bus to a distant town to visit my grandparents. Since the journey lasted several hours, the bus stopped in a small town for 30 mins so that passengers could use the facilities and get some lunch. It was a small town in India. While the bus was parked in the bus station, a lady selling locally grown tomatoes showed up. She had a basketful of fresh looking ripe tomatoes and was hawking them as good quality produce right from a local field nearby. Price was some 50% off compared to what you will normally pay in a regular grocery store. So, several passengers including my parents bought some. Within minutes that lady managed to sell the whole basket to various bus passengers, thanked us for brisk business and left. Later that day when we reached my grandparents' home, my mom started chopping up the tomatoes only to realize each tomato was pumped with a lot of water using a syringe to make them weigh a lot more. Not only was that an annoyance but the increase in weight more than cancelled out any discount we thought we got in price. Starting from smalltime cheatings of this kind, billion-dollar frauds like what Bernie Madoff pulled off few years back, there are a lot of fraudsters in the world. You might have even seen some of your friends and relatives describing them as clever, as long as they are not personally affected.

In the last chapter we saw that when there is no need for recurring interactions, from an individual's point of view, cheating as much as you can to get the maximum advantage for yourself might appear to be the best strategy in the world. This was borne out clearly through the Prisoner's Dilemma thought experiment. The lady who sold us those water filled tomatoes knew well that the customers will realize her fraud as soon as they try to use her product. But she didn't care since she was dealing with one-time customers that will probably never see her again. So, it doesn't matter if they get angry at her. She is not worried about retaining them as her long-term customers. So, she is better of making as much money as possible. There will always be a fresh set of suckers on the bus tomorrow.

Alternately if you are running a business in a town selling vegetables to repeat customers, this trick won't work. In that circumstance it is even possible that while you cannot afford to cheat your customer to get an advantage, your customer might try to cheat you. For an example, a well-dressed respectable looking gentleman may show up in your store, buy a bunch of stuff, and then tell you that he will send money with his driver sitting in the car outside and run away never to return. Since he is not planning to keep interacting with you repeatedly in future, he doesn't care if he really upsets you. In that case, does it mean that trying to cheat as many people as you can to gain as much advantage for yourself is the best strategy in life to get ahead? If everyone tries to cheat everyone else, how can that society and its economic system of exchanges thrive? Are all the people in the world who trust others suckers? If the answer is yes, what about all the teachings we received from our parents, teachers and elders that we should be kind to others, need to help even strangers as much as we can and what not? Is that all nonsense?

Iterative Prisoner's Dilemma

Professor Robert Axelrod of the University of Michigan wondered about this problem. Setting aside the moral values we are taught by the society and religion, he wanted to investigate this problem more objectively with a very clear measurement process and decide what approach is better. To carry this out as a proper research project, he announced a competition around the year 1980. It was based on the same Prisoner's Dilemma problem we discussed before.

Robert Axelrod

Just like the Golden Balls TV show, the two participants have to choose either “Split” or “Steal”. Let us say Alan and Bryan we encounter in the last chapter play this round. As we discussed before, depending upon their choice, the situation will result in one of the four squares. As per the rules of this game, if both had decided to cooperate and pick “Split”, each gets 3 points.

When one picks “Split”, if the other tries to cheat and pick “Steal”, the one who cheats will get 5 points. But the one who picked “Split” will not get any points.

If both attempt to cheat the other and choose “Steal”, both get just 1 point. These four possibilities are depicted in the figure below.

		Bryan: Split	Bryan: Steal
Alan	Split	A : 3 B : 3	A : 0 B : 5
	Steal	A : 5 B : 0	A : 1 B : 1

Though till now the steps and rules resemble The Golden Balls game show, it doesn't stop with one interaction between the two opponents. Learning what the outcome was, they have to play again, and again and again, a total of two hundred times. Since the contestant plays with the same opponent over and over, one can assess how reliable or honorable the opponent is and adjust the strategy from play to play. If you are the player, you can choose to "Split" or "Steal" depending upon your history with the opponent over the past rounds. If the opponent is a bad one trying to take advantage of you in each round, you may decide to hit back choosing "Steal" repeatedly. This is equivalent to not buying tomatoes from that lady who sold you tomatoes filled with water or even reporting her to the police for fraud.

Points awarded are intentionally designed to be very high to the person who cheats and very low for the person who happens to be trusting. When both contestants cooperate, both are rewarded reasonably well and when both decide to cheat both are punished with very little or no reward. The rewards structure must penalize the trusting contestant when the other cheats. This is important. Since the end goal is to collect as many points as possible, this reward structure will clearly verify if even though cheating in the individual rounds is advantageous, does the strategy to cooperate work in the end while the process is iterative.

Once he designed these rules, Professor Axelrod asked several experts in various fields working all over the place to send in their strategies to compete and win in this game. He received fourteen submissions. He converted each submitted strategy into a computer program. He then let each strategy compete with the remaining thirteen strategies two hundred times each. If we think about this process, it can be considered equivalent to a small society of fourteen people where each one of those fourteen interacts or trades with the remaining thirteen people in the society two hundred times each.

So many different types of strategies were enthusiastically submitted for evaluation. These are some examples:

- Always cheat your opponent (i.e. always choose “Steal”)
- Always trust your opponents (i.e. always choose “Split”)
- Keep switching between “Split” and “Steal” in successive rounds.
- Keep choosing “Steal” until one hundred points are gathered. Then start choosing “Split”. (This is equivalent to aggressively accumulating wealth in one’s life

initially and then once we have enough, switching over to being a philanthropist.

- For the first fifty iterations gather what the opponent does. If most of the time opponent hits you with “Steal”, you hit back with “Steal”. If the opponent was kind to you most of the time, then be kind and choose “Split”.

Spend some time to think through what strategy you will adopt to be successful in this evaluation. Perhaps write it down to be

Anatol Rapoport

sure.

Since we are going to play the game iteratively with the same opponent for two hundred times, we can assess profit & loss, learn our lessons and develop our strategy in so many different ways. So, several experts from the fields of Math, Economics and Social Sciences that participated in the competition had proposed detailed, complicated strategies they thought will be required to win the game. At the end of Axelrod’s evaluation, can you guess what strategy turned out to be the winner? It is a simple one named “Tit for Tat” submitted by an Austrian Math Professor named Anatol Rapoport. As per this strategy, the program will presume the opponent is trustworthy and choose "Split" during the first iteration. If the opponent also chooses "Split", it will repeat the same choice the next time around as well. If the status quo continues, both sides will continue to choose "Split" repeatedly, bagging three points each for every round.

When Rapoport's algorithm chooses "Split" on the first round, if the opponent chooses "Steal", in round two, Rapoport's algorithm will hit the opponent back with "Steal" to punish for previous round's misbehavior. If the opponent chooses "Split" the second time, turning out to be good & trustworthy, in the following round, Rapoport's algorithm will quickly forgive the opponent and choose "Split" again. In other words, Rapoport's strategy is to trust the opponent and choosing "Split" giving the benefit of doubt. From round two onwards, it will simply imitate whatever choice the opponent made in the previous round.

We can equate this strategy to trusting a stranger you meet in real life and interacting with the stranger nicely giving him/her the benefit of doubt and then depending upon how the stranger behaved during the first interaction, reciprocating in kind during the next interaction with the opponent. When we realize someone has cheated us in real life, we may want to hit back to teach that person a lesson. But then it may not be a good idea to keep that person permanently in the baddie column. Instead we may want to assess how he/she is behaving in subsequent interaction and if they changed their ways behaving nicely, without any ego or permanent ill will towards that person, we can also start cooperating. Not retaining a long tail memory of what happened eons back, makes it easy to cooperate with people who are willing to cooperate with us while retaliating people who are not nice to us is the essence of this strategy codified with a very simple "Tit for Tat" approach.

Since what this strategy does is easy to comprehend, more elaborate schemes understand this approach and start cooperating with it quickly. Even if both contestants use the same strategy, both start racking up points and so it works out well for both sides. This approach doesn't worry about how many points the opponent has gathered already, how many times we got cheated in the past iterations or any such factors. Thus, this is equivalent to playing the game without the notions of jealousy, greediness or revenge clouding up one's judgement. At the same time, this strategy doesn't hesitate to hit back when cheated. So, when all the fourteen submitted strategies were made to play against each other for two hundred iterations each, this simple "Tit for Tat" strategy is the one that won the game with maximum points in the end. Strategies that attempted to cheat everyone as much as possible received only minimal points and ended up pretty low in the ranking.

Few years after announcing the results of this competition, Professor Axelrod ran the competition a second time. Since the experts who submitted the strategies for the first competition knew the results and how their approach performed in the competition, many redesigned their strategy with lot more subtlety to cheat the opponent gathering advantages to themselves. With increased interest and attention to the event, Professor Axelrod received thirty submissions this time. Despite the improvements others added and the increased number of submissions, in the end the simple "Tit for Tat" strategy won the competition again.

If twenty-nine out of the thirty submissions use cheating strategies with just this one entry uses "Tit for Tat", it may not win the game. But with repeated trials, it has been shown that strategies that aggressively try to cheat and take advantage of the opponent tend to kill off each other, while the "Tit for Tat" type strategies that tend to give the opponent the benefit of doubt, tend to rise up the ranking when the game is played iteratively for multiple rounds.

When these results were well understood, researchers investigated further to see if this approach can be improved even more. Consider the scenario where Alan and Bryan are playing this iterative game. In one iteration, let us say Alan trusted Bryan and chose "Split" while Bryan cheated Alan choosing "Steal". Since they both are using "Tit for Tat" strategy, in the subsequent iteration, both will switch their choices mimicking their opponent's choice in the previous iteration. This will result in Alan choosing "Steal" and Bryan choosing "Split". This will force both of them back to their previous choice in the following iteration. When they reach this stage, they will be stuck, repeatedly switching their choices and not being able to get out of this choice sequence. To be able to recover from such situations, more altruistic strategies where you trust the opponent twice were introduced. In this model, you will trust your opponent on the first interaction and choose "Split" to be kind. Even if the opponent hits you with a "Steal" on that round, you will remain magnanimous and choose "Split" one more time. If the opponent again chooses "Steal", you switch your selection the next time. But if the opponent choses "Split" the second time, you continue to choose "Split" in subsequent iteration as well. Such a strategy may help Alan and Bryan who inadvertently get into the rut we stated above get out of it. Since even this level of kindness can be taken advantage of by aggressive algorithms (by cheating twice to get the advantage and then cooperating to gain trust only to cheat twice again repeatedly), minor tweaks to this kindness have also been introduced. Thus, for example, without being kind two times in a row predictably, using "Tit for Tat" most of the time but once in a while unexpectedly being kind twice is one such strategy.

Deer Hunting

Imagine two hunters going out to hunt. If they decide to cooperate with each other, they can potentially manage to capture a larger animal like a deer. But if they go their separate ways and hunt individually, scoring a large animal may be more difficult. Thus, they may each end up getting just a rabbit or two. Capturing a deer is much more profitable compared to scoring a rabbit. But if you are not sure whether your partner is a skilled hunter or not, you might end up working with the partner for a long time and then return home empty handed since the partner was not as talented as you initially imagined. Compared to that scenario, you working alone and at least bagging a rabbit might be better. So, if you are not sure how skilled or trustworthy your partner is, does it make sense to cooperate for a better return or go your own way ensuring a definite but smaller success? This is also a variation of the same Prisoner's Dilemma thought experiment. In the world of commerce, software giant Microsoft and semiconductor giant Intel joining hands to bring the Wintel Revolution to reality can be viewed as two hunters cooperating to score a fantastic win. If they had not worked together, neither the companies nor the world would have experienced the PC revolution of the 1990's. Being good companies, they probably would have been successful but at a much smaller scale, equivalent to capturing small rabbits.

We learn a few unintuitive lessons from these series of thought experiments and related research projects pertinent to commerce. Though it may appear advantageous to cheat your opponent to gain something in the short term, this approach may work only on single interactions. Since we increasingly live in societies that can track an individual's behavior through technology very well, such approach will not work in the long run. Think of even the lady that sold us water filled tomatoes decades back in a bus stop. If she does the same thing now, it is easy for the passengers to post notes on websites describing what happened to them on that bus stand. They may even be able to post a picture of the women taken either using their smartphones or from some security camera nearby. This will prevent that lady from repeating that trick on that bus terminal day after day. Idea can be extended by an order of magnitude to bigger entities like a shop or much larger corporations. Be it in the society or in the jungle where different organisms compete through Darwinian rules of evolution, those who cooperate may thrive as a society and do well in the long run. This was borne out by a very interesting analysis in the field of commerce. US had managed to develop a culture where you can hire someone from the street to work in your organization and the chances are the new employee will be trustworthy and won't steal from your company. This is not true in most other countries in the world, including many Western European countries that are industrialized and advanced. As a result, when a small company in US has an opportunity to grow, they quickly hire employees and start exploiting the opportunity in front of them. But in countries where a person on the street may not be trustworthy, smaller companies hesitate to hire strangers to work in their company. So, many mom and pop institutions like a restaurant or a bakery grow only to the point where they manage to employ everyone they can from their own extended family and then hit a plateau. The opportunity and the needed company infrastructure really must grow for them to be able to go and hire lot more employees from the society they live in to grow further. This is apparently holding back a lot of small companies in dozens and dozens of countries from growing into their full potential.

An extension of this notion is explored in another thought experiment model proposed to make better use of land available in a country. It is easy to see that the available land in any land is not being used well. There can be large swaths owned by individual or corporate landowners that are underutilized while there are other entities that could better utilize the land who are not able to get access to available lands. In the proposed model, all the land available in the country will always be on sale. Current owner of the land will determine an acceptable price they are willing to take to give away the land. If any person/entity is willing to pay that price, then the current landowner is obligated to vacate the land and hand it over to the new owner after accepting the agreed price. As long as no one is buying the land, the current owner gets to enjoy the land while paying a tax that is determined by the price the owner is willing to accept to sell the land. Just to make it easy, let us say the landowner must pay 10% tax each year. So, if the land is worth $100,000, the owner pays $10,000 in tax each year. But anyone can pay $100,000 to the owner and buy the land anytime. The owner trying to protect the land by setting too high a price, like say $1 million, will be forced to pay $100,000 in taxes each year. But to cut down the tax, if the landowner sets the price too low, say $10,000 which is far lower than the market price, next day someone will step in to buy the land from the owner. Thus, this model will ensure that all the available land is owned by those who really value it and so will derive maximum enjoyment (by paying high taxes) or maximum benefit (by putting the land to work in the most profitable way). This idea can be extended to other resources such as homes or factories or automobiles also, eliminating the trustworthiness playing havoc with the economy and letting market forces determine the true worth of any given item, thus putting it to best possible use on the best possible hands.

As such these understandings, results of the research efforts will apply to various organizations, departments, environments, life forms and so on where interaction between different entities is required for them to thrive.

Lesson from a Flipped World

Since the field of commerce is closely related to the field of economics, though this book has several chapters, this and the last chapter can be said to have formed a tight knit subsection. Since the beginning of the last chapter started with a professor-student story, we might as well bookend it now with another professor-students story.

Few years back this story actually occurred in the Johns Hopkins University located in the U.S. Peter Brolick is a Computer Science Professor in the University. For a class he was teaching he decided to make the grading relative. So, rather than individually grade papers, he announced that whoever does the best in the final exam will get the maximum points, which is A grade as per American University grading systems. Then based on that standard, remaining answer sheets will be graded. His thinking was this will push each student to strive for that top grade.

You know what the students did? After discussing amongst themselves, everyone decided to boycott the final exam. On the day of the final exam, Professor Brolick showed up to the exam hall with the question paper. But all the students who took the course, stood outside without entering the classroom to take the test, ensuring that no one else did either. If anyone entered the classroom to take the test, it would automatically force all the other students to take the exam as well. If anyone had gone in and simply returned the answer sheet after writing his or her name in it, that would have become the best answer sheet as per professor's definition. But since no one took the test, it turned out that everyone's answer sheet was as good as the best answer sheet. This forced the Professor to give an A grade to every student in the class.

When asked about this incident, Professor Brolick said that he knew well that Johns Hopkins students were very intelligent, and they do fight for good grades, but he never expected them to come up with this scheme and stick to it as a whole class. Naturally he stopped following this relative grading mode ever since. So, students in the subsequent semester couldn't pull off this stunt to get an easy A. But you wonder whether to admire the students that thought about this scheme on that first and only round and managed to pull it off without anyone in the class betraying the rest or appreciate the professor who didn't complain about the students' approach, didn't renege on his promise but gave an A to all the participants. Stepping back, apparently some researchers are still trying to determine whether this is a proof for the saying, "United we stand, divided we fall" or evidence that a whole bunch of people scheming together in a society can make a fool of even an erudite scholar.

--------- o0o ----------

Medicinal World

The aim of medicine is to prevent disease and prolong life, the ideal of medicine is to eliminate the need of a physician.

– William J. Mayo

Adam and Claire traveling together get into an automobile accident. They are able to place a call to the emergency service from their cell phones and are picked up by an ambulance that takes them to a nearby hospital. It is late night and the little hospital located in a small town has just one doctor on the premises. The doctor examines both promptly and determines that Adam is seriously injured and if the doctor doesn't perform an emergency surgery, he will lose both his legs. Claire has only a small fracture in one of her fingers. The doctor may just have to buddy tape it to keep it immobilized so that it will heal itself. So, he figures that he can attend to Claire later and starts working on Adam prioritizing his treatment. But Adam blocks him saying, "I understand that I am badly hurt. But Claire is my daughter. Her wellbeing is more important to me than my own legs. So, please go and take care of her finger first." If you are the doctor on duty, what will you do?

To complicate the situation even further, let us say the accident they got into was caused by their car hitting another car. Bill who was traveling in the second car is also hurt. Bill's health condition is precariously hanging between Adam & Claire's conditions. In other words, if the doctor does not start treating him immediately, he will lose one leg. Bill is also in the same hospital room with Adam & Claire. Having examined all three patients, the doctor moves in to treat Adam since he needs the treatment the most. When he blocks the doctor asking him to treat Claire first, the doctor says, "You have the right to refuse treatment. That is your prerogative. But you can't dictate who should be treated next" and moves over to treat Bill. In this A, B, C choice problem, though she is not hurt badly, the little girl Claire is screaming at the top of her lungs demanding attention. Since you are the only doctor in the hospital, Bill looks at you and says, "Adam has said he doesn't want treatment. Though Claire is a little girl crying, since her injury is a minor one, and mine is pretty severe, it is not appropriate for you to treat anyone other than myself." As the attending physician, what is your responsibility? This thought experiment question can be made more complicated and confusing easily. Let us swap Claire's character from being a little girl crying for attention to a world-renowned piano or violin virtuoso. So, though the injury she had is a comparatively minor one, if it is not treated promptly, the whole world may lose a fantastic musician and her creativity. Under this scenario, does the priority of who should get treated first change?

You are not allowed to answer this question with simple answers like, "I will ask the nurse in the hospital to assuage the child for a while until I am done with more urgent tasks." The problem we are trying to evaluate or potentially resolve is a serious one. When there is a shortage of medical resources, be it physicians or lifesaving drugs or surgery equipment, how do you ration it amongst multiple deserving people who are in need? This is the question being analyzed. Be it a large plane/train related accident or massive natural disasters like a flood or earthquake, the need for rationing available care will pop-up its head. Personnel working in the field will be forced to manage the situation with what they have. So, it is easy to understand that this thought experiment has very real life & death practical consequences.

In a warzone, when multiple soldiers & civilians get torn up by fired bullets or exploding bombs, field medical personnel and people that belong to organizations like *Doctors Without Borders* are routinely put under the daunting dilemma of having to play favorites so that they help victims survive. But a recent disaster has taught us how important and close such decisions are to any part of the world.

You may remember Hurricane Katrina that devastated the southern coast of the US in the year 2005. The state of Louisiana was the worst affected. Though the meteorologists had been predicting a powerful storm for days, since the preparatory as well as the recovery work needed to face such a powerful Category 5 storm wasn't adequate both at the state and the federal government level, thousands died and tens of thousands struggled to recover. This debacle is well known legend now. In subsequent years countries like Haiti and Pakistan faced natural disasters (earthquake and flood, respectively) resulting in similar hardships to their citizens. When their respective government officials were criticized for not delivering the needed services, they routinely brushed aside such criticism saying, "Don't you know how badly the US, such an advanced rich nation, delivered services to their citizens during Katrina? We are doing much better than them despite being underequipped and poor." It left such a bad stain in the US's record and capability for facing such natural disasters.

By now several books and documentaries have been produced describing how quickly things went from bad to worse all over that region at that time. But what specifically happened in a hospital known as Memorial Medical Center located in the city of New Orleans maps to our current thought experiment precisely.

If you logon to Google maps and zoom in to inspect the southern coast of the state of Louisiana, you will quickly realize that it doesn't have a clean coastline but has what appears to be frayed edges of a rag cloth or a torn up beehive that is lying in water completely mixed up. I lived in the city of Baton Rouge in Louisiana for twelve years. Since Baton Rouge and New Orleans are only sixty miles apart, I am quite familiar with the geography of that area. That whole area is full of swamps and marsh lands. New Orleans, the most populous city in Louisiana, is actually sitting a few feet below the sea level. If you start digging the land, often you will encounter the water table fairly quickly. So, in fact you can see the dead being "buried" above the ground level. To prevent water from entering the city, decades back engineers had constructed a series of levees. The city is completely dependent on these levees. If they ever breach, water will fill up the city as if it is a basin waiting to receive incoming water.

Memorial Medical Center is an eight-floor building constructed in 1926. It's been standing there in harsh sun, extreme humidity, rains and hurricanes for eight decades serving the local community. So, in times of natural disasters it was always thought of as a reliable friend willing to lend a shoulder to lean on. The day Katrina hit the shores, August 29, 2005, was a Monday. The ferocity of the hurricane promptly cut the power off pretty much for the entire city. The MMC facility had backup generators in place and they promptly came on keeping the lights in the hospital lit that day. So, Monday wasn't bad at all from the hospital functionality point of view.

Since New Orleans is located below sea level, with abundance of caution and good planning, MMC had backup generators installed two floors above the ground on the second floor of the building. When the levees built around the city breached Tuesday, the city started quickly submerging in water. MMC was caring for about two hundred and fifty resident patients. The care providers such as doctors, nurses, and support staff added to this total. But since it was a long-standing well-known building in the area with power, many more locals sought refuge in the building, burgeoning its population close to two thousand that day.

As the city was starting to get flooded, administrators of the MMC figured that the situation may start to deteriorate seriously within a day or two. So, they started acting swiftly trying to evacuate the vulnerable patients first. Since roads and the hospitals surroundings were flooded, the best possible approach was air transport. They listed out the ICU (Intensive Care Unit) and pediatric ward patients, many of whom were on the seventh floor. Staff started bringing them down via elevators from the floors they were in and took them to the part of the facility that hosted a helipad on its top floor. Then again using the elevators, they took these patients up to the helipad and shipped them off via helicopters to other hospitals in nearby towns that could care for them. As the sun started setting down on Tuesday, one by one about sixty such patients had been transferred out of the facility. Till this point in time, we can say that the administrators, physicians and care providers in a big city located in a rich industrialized nation handled the disaster as you would expect them to handle it. It was orderly, well thought out and proceeded as planned. Guess the cruel fate didn't like these orderly proceedings and so decided to conspire against the hospital. Though the generators were set up on the second floor of the hospital, some of the main breakers supplying power for the whole facility were located on the first floor. As the water level started rising throughout Tuesday night and into Wednesday morning, the entire first floor of the facility went under water thus crippling the breakers controlling the power lines. This resulted in all the backup generators shutting down.

The ICU was still caring for nine patients who were breathing through ventilators that depended on electrical power.

With power totally cut off, their condition immediately became precarious while all the elevators in the building simultaneously came to a grinding halt.

Undaunted, residents of MMC looked for alternatives to transfer out the remaining patients. Luckily there was a parking lot in the facility that was close to the floor where the patients were located. The shortest way to the parking lot from the floor was through a three-square foot hole that existed to help transport supplies in and out of the floor. With everyone pitching in, patients were loaded one at a time on stretchers and carried over by hand to the adjacent parking lot through that three-square foot hole. Then they were loaded in available cars that took them to the fifth-floor garage. There was a small unused stairwell on that fifth floor connecting it to another part of the top floor where potentially helicopters could land. Since those stairs weren't designed wide enough to carry stretchers, two volunteers at a time helped each patient slowly climb through those stairs to reach the top floor to board the helicopter. By then it was late night. So, the pitch darkness and the small beams of flashlight in an atmosphere of heat and humidity started to make the task very difficult. As the volunteers and the staff were putting in heroic efforts to get these patients transferred out, five of those nine patients died.

Though this was starting to look like a frustrating setback, everyone in the facility thought all they had to do is manage the situation for a few more hours till Wednesday morning. Once the sun rose, they expected military and coast guards to show up with a dozen helicopters and transfer everyone out in a jiffy. There were fewer than two hundred patients left. Should be a piece of cake for the military professionals well trained in rescue efforts. Right?

The sun rose on that hot and humid Wednesday morning. Though the residents of the hospital needed help, the surrounding areas were in dire straits. One could see entire panic-stricken families standing on roof tops and even on top of cars, waving for anyone to help them. Neither the military nor the city administration were prepared to handle this level of flooding. The available helicopters became a precious resource that had to be used judiciously to maximize their utility. When the military concentrated on saving people on treetops and car tops threatened by imminent drowning in the next hour or two, there were no helicopters left to save the remaining people in the hospital.

The hospital staff figured that they must help save the remaining patients when there is sun light using any small boats or helicopters as and when they became available infrequently. Since speed was of the essence, the physicians on duty decided to sort the remaining patients into different groups. Whoever had Group 1 stickers on had to be saved first. People with Group 2 stickers were definitely of lower priority compared to Group 1 patients. Group 3 patients will be helped only after all the Group 1 and Group 2 patients have been helped/evacuated. Since the arrival of helicopters and boats was infrequent and rare, classifying patients into different groups and prepping them appropriately makes complete logical sense. This will certainly utilize the available lifesaving opportunities very well. But then the question of who belongs to Group 1 and who belongs to Groups 2 & 3 pops up. If you are the deciding authority, how do you go around making this decision as it will potentially result in life or death? Should we place the patients in the worst health condition in Group 1? Or perhaps the patients in the best health condition in Group 1 so that they have the high probability of living for a long time to serve society later? If you decide to place someone in Group 3, how would you justify that assessment to that person or to their family?

Anna Pou

Dr. Anna Pou was one of the physicians serving in the hospital during this crisis. She was a surgeon who specialized in head & neck related surgeries. Being forced to categorize the patients, she decided to place all the patients who had a DNR (Do Not Resuscitate) order in place as well as those who had a pretty low chance of long-term survival in Group 3. In other words, she was effectively giving up on these patients, deciding that it is not worth spending the precious lifesaving opportunities on them. She placed ordinary patients who were suffering from some routine ailment in Group 2 so that they could be sent away as and when helicopter slots became available. Those who were in the best physical shape were classified into Group 1 and recommended shipping them off via boats as and when they came by.

As each minute was grinding away ever so slowly, the situation was only getting worse. By Wednesday afternoon, without the comfort of air conditioning, sunny weather outside raised the temperature inside the building past one hundred degrees Fahrenheit. Louisiana is well known for its extremely humid climate in the summer months. The hospital administrators foresaw the possibility that the staff may have to stay in the hospital for more than twenty-four hours continuously as the storm battered the city. So, they allowed the staff to bring their pets with them as there may be no one at home to take care of them for a whole day or more. This is a routine exemption given to staff whenever there is the possibility of them staying in the building for prolonged periods of time. The cats and dogs the staff had brought to the facility were running out of clean water to drink and were also suffering from the excessively humid climate and rising temperature inside the building. Some pets had started going into caesuras and fits. The owners of those cats and dogs who couldn't stand to see them suffer, started crying and requested they be put to sleep. Physicians had to proceed with the mercy killing process as there seemed to be no other option at that point.

It was Thursday morning. TVs were not working as there was no power. So, the staff isolated in the facility listened to radio stations that were broadcasting messages that were often totally incorrect. Some police officer in the city spoke to a radio station saying he saw something that looked like a shark swimming around a hotel in the city that had submerged in water. Who knew when it may swim into a building and start eating people?! Someone else reported that he saw military personnel parading around with curfew in place allowing the military personnel to shoot people who didn't obey the curfew. There were unconfirmed news reports of malls and stores being ransacked by hooligans and gang members since there was no police presence to prevent them. Messages of this kind heard on the radio started rattling people. Though many such hearsay reports turned out to be nonsense, there were indeed atrocities on going in the city. The Superdome is a very large indoor American football stadium that is the home of the local American football team named New Orleans Saints. That facility was used to house thousands of the city's homeless residents. Riots were breaking out amongst those residents, many were attacked, several women were raped and a few actually got killed.

By Thursday many of the MMC medical staff were reaching the stage of nervous breakdown. They were crying loudly, starting to pray and babble believing that their lives may be at the end. There were about a hundred Group 3 patients (i.e. those who were pretty much left to die) made to rest in the corridors of the facility. With rising levels of noise, stink, temperature and humidity combined with inadequate food and water, they were all starting to faint and yell/cry loudly seeking attention. All this was happening in the middle of New Orleans, a major American city that was functioning perfectly like a first world city just seventy-two hours before.

Since the temperature remained high with no sign of helicopters, this sad situation continued throughout Thursday. Dr. King who was working in the facility that day claims that he saw Dr. Pou walking around with a collection of loaded syringes in her hand telling those patients, "I am going to give you something to ease to your pain" and then injecting them with something. This is not the typical treatment methodology in which physicians take care of patients. So, Dr. King was effectively accusing Dr. Pou of killing several patients quietly on that day.

It took a few more days for the situation to improve. Eventually when things came under control and the cleanup effort started in earnest, a total of forty-five dead bodies were found inside the MMC facility. In the post-mortem process, twenty-four of those forty-five patients were determined to have excessive amounts of morphine in their system that could have killed them.

A year after that natural disaster called Katrina ruined New Orleans, the state of Louisiana conducted multiple investigations to determine if Dr. Pou murdered anyone on that fateful day. She was arrested and the state ran a grand jury investigation. But the prosecutors couldn't put together a convincing set of evidence to prove the case and so it was dropped in the end. Subsequently when TV interviewers posed questions like, "Did you commit mercy killing to let go patients who were really suffering that day?" Dr. Pou vehemently denied doing any such thing. But even beyond Dr. Pou's conduct, there were other physicians who were forced to address patients who were really suffering. Dr. Cook is another physician who was there that day. He had provided statements under oath claiming that he indeed asked a nurse to provide a large dose of morphine to an elderly cancer patient who was suffering badly in pain. He argued that if she had died due to that level of morphine administration, he still considers it the right decision.

There is this concept called *Double Effect Doctrine* in the field of medicine. Even if a drug provided by a physician could have killed the patient, if at the time of dispensation the physician had provided that medication to address an ailment, with patient's care being the main reason, then the Doctrine says that the physician is in the clear, in other words the prescription should be considered perfectly reasonable. If you evaluate Dr. Cook's actions by that standard, it is hard to accuse him.

Even a decade later, some of the staff who were working there on those few days are apparently suffering from psychological issues and are struggling with PTSD (Post Traumatic Stress Disorder) type problems. Analysis of those few days and the lingering aftereffects that still persist in society has led researchers to ask the question, "Long before medical professionals face such daunting decisions, can the society we live in evaluate such dilemmas and provide a priori guidance as to what should be done under such circumstances?" This may relieve the hospital staff from having to make such horrible decisions under stress.

To help resolve this predicament, a few years back the state government of Maryland invited a cross section of the general public to participate in a debate. On a Saturday morning, several hundred volunteers gathered in a large conference hall to hash this out through a thought experiment. In the scenario described to them, a virulent strain of a new type of flu is sweeping through the world, including the U.S. Since the communities were totally unprepared, many citizens die. As it hits Baltimore, the city realizes that there aren't enough ventilators to help all the victims struggling to breathe. Since available ventilators are a precious resource that will determine who will live to see another day and who will die, volunteers have to discuss and decide who in the community should be given priority to get access to the available set of ventilators. Participants analyzed the problem in front of them from multiple perspectives. Thus, for example, if there is a hapless young man in town who is a habitual alcoholic and a petty thief, should he be given priority because he is young and so once saved, he will remain alive for a long time? What if he is competing against an elderly priest in the local church? Should we allocate all the ventilators to children and infants first? What if that results in all the infants staying alive with not enough adults to take care of them that results in the infants' eventual death? Should all the medical staff get priority so that it will provide enough qualified people to treat the rest of the society? Should the citizens of the country be prioritized over non-citizens? Or under such extenuating circumstances, should national boundaries not matter anymore? What if the non-citizen happens to be a more productive member of the society? Under whatever protocol that is in place, if a person who received a ventilator seems to be still struggling for life, should we give up on that person and reallocate that ventilator to another patient who may benefit from the care? Does the priority increase if that other person is young (so that they will live long and be helpful to the society) or old (since they can't take care of themselves)?

If some well-defined protocol that is agreed by everyone is in place, at the least everyone in the community may know and understand what is expected to happen under these disasters. Isn't that better than not having any protocol that results in massive chaos and free-for-all fights? Particularly under such stressful situations when emotions will run high, isn't it important to maintain law & order? After hours and hours of discussion, this is the series of conclusions made by the participants.

1. When you take a closer look, we realize that no priority scheme will appear correct under all circumstances. So, the best protocol should be one that will extend the total lifetime of all the human beings under the system.

2. Other factors such as race, religion or nationality shouldn't factor into the priority-making decisions.

3. People who are really sick and so won't benefit from the treatment in the long term should be deprioritized. So, though it may appear very unfair and cruel, those with failing kidneys, severe heart conditions, and cancer-type diseases that do not provide a good long-term prognosis should all be placed very low on the priority list.

4. If one must be chosen from a set of say four people who all have equal claim to that scarce resource, then the one candidate should be selected via a random lottery process.

When you read through the collection of these decisions, as disturbing as they may be, we realize that these are not necessarily the kind of decisions that will be made in a far-off place and situations we will never encounter. The COVID-19 pandemic practically brought these considerations to our doorstep. US Government reported that there may be a need for millions of ventilators, while the stockpiled reserves were only in the range of 10,000 units. If social distancing and other measures work to flatten the curve during such pandemics, we may get away with it. If not, real life and death decisions may need to be made to distribute scare resources such as ventilators correctly. There are proposals to use lottery systems to decide who gets what. There are also considerations to setup panels to make such decisions in some states making people extremely uncomfortable about joining such panels. Leaving aside pandemics that exacerbates the situation to extremes, let us just consider kidney transplants. There are thousands of patients all over the world who suffer from failing kidneys. But the number of kidneys available for transplant surgery is abysmally low. So, when a kidney becomes available, do we give it to an elderly person who has been on the waiting list for a long time patiently suffering through the ailment or to a younger patient who got on the list only recently but will live a longer life benefiting more from this rare resource? This is a question that gets debated every month in every country.

To analyze the kidney assignment issue alone, a thought experiment named Kidney Club was proposed. Instead of waiting till our kidney stops working, let us say we all get examined by medical professionals and determine who will be a good donor match for us, if we ever need a kidney. If you and I are found to be good matches for each other, as per this thought experiment, we will sign an agreement. As per the agreement, if anytime during the rest of my life both of my kidneys fail, you must donate one of your kidneys to me. By the same token, if both of your kidneys fail at some point, I will donate one of my kidneys to you. Sounds like a reasonable plan and a good idea. Right? The purpose of the thought experiment is to analyze whether such agreements should be allowed.

Though it may sound like a good idea at first, such agreements are not legally enforceable. Let us say four years after you and I signed such an agreement, both of your kidneys indeed fail and so you are asking me to donate one of mine to you for transplant. Since I wasn't really expecting this situation, I may simply tell you that I no longer want to be on this contract and so I am cancelling the agreement we signed before unilaterally. What can you do? You can't complain to the police and expect the legal system to enforce the agreement by forcibly removing one of my kidneys. For all you know, I might have signed that agreements four years back in good faith. But perhaps after going through the surgery details and the risks I am taking on, I might have become genuinely scared and so may want out of the agreement. In that case, the government forcing me on to the table to rip out one of my kidneys when I don't want to go through the procedure crosses over into the domain of human rights violation and/or inhumane treatment of a patient that cannot be allowed. So, no country on this planet supports such contracts.

As we continue to think through such plausibilities, we realize that these kinds of medicine-related decisions that involve life & death should all be ideally made by nature or your guardian angels or God or some such equivalent as we will all be extremely uncomfortable making them ourselves as rational human beings. As much as I have thought about it, the best approach to handle this problem has two steps to it:

Step 1: Setting aside our personal biases and preferences, rationally figure out and write down how such decisions should be made so that we have an agreed upon plan amongst us.

Step 2: Pray that we never have to invoke and exercise such a plan anytime in our lives.

--------- o0o ----------

Democratic Planet

Shallow understanding from people of good will is more frustrating than absolute misunderstanding from people of ill will.

– Martin Luther King, Jr.

The concept of democracy and the idea of setting up and running a government based on that concept were gifts given to the world by the Greeks. The same old Greek civilization also provided another great gift to the world in the form of Western philosophy. But some of the iconic Greek philosophers such as Socrates and Plato thought that this model of democratic governance as we practice today would never work well in

Socrates

reality.

To explain why he disapproves of this model of democracy, Plato proposes a simple thought experiment as if it was posed by his teacher Socrates. You may know about the Socratic Method of teaching, where instead of giving the students the question and its answer, the teacher simply asks questions, prompting the students to answer them and in the process leading the students to find the answers themselves. Using that model, in this TE, a doctor in a town examines a patient suffering from an ailment and prescribes a medication. The medication happens to taste bitter. Noticing this simple fact, a sweet seller standing nearby tells the patient, "Man..this stuff is so bitter. Why do you need to tolerate this nonsense listening to this guy? Come to my store. I can sell you tons and tons of mouthwatering sweets. You can eat all you want and have a ton of fun." Who will the patient choose to go with is the question Plato poses through Socrates.

You may wonder what kind of a silly thought experiment is this, since anyone knows that however bitter it may be, you need to take the medication a doctor prescribes so that you get cured of the ailment. But for you to make that correct choice, you should first be aware that you have an ailment and the doctor is indeed trying to cure you by providing appropriate medication. If you don't have that level of understanding and tend to behave like a small child, you may complain about the bitter medication and insist on some sweet fun things to eat. At the least if you are clearly suffering from the ailment with obvious symptoms, others you trust may try to explain the rationale behind taking the medication irrespective of how it tastes. What if the disease you have is of a kind that doesn't result in any obvious discomfort or external symptoms? We know that most cancer incidences are caught early these days thanks to extensive scanning and testing procedures we have at our disposal. As a result, they do get detected a lot sooner than any physical symptoms or discomfort shows up in the patient's body. In those situations, when the cancer patient is not feeling any pain or discomfort, wouldn't the sweets offered by the vendor sound more appealing than the cancer medication a doctor prescribes?

Plato & Socrates were not necessarily against democracy to promote some sort of archaic monarchy as the right form of government. They were in support of democracy but wanted the people with voting rights to know what they are doing. If the patient mentioned in the thought experiment above is reasonably knowledgeable, he will understand that though the medication provided by the Doctor tastes yucky, since it can cure his ailment, he should take it as per the Doctor's advice. Similarly, he may know that the sweets offered by the vendor may taste really nice now but will not cure his ailment and may even ruin his health further, which means he should not consume it in place of the medication. Plato & Socrates wanted people with such common sense to be the ones with voting rights. They fully expected politicians of questionable character to step forward with promises that are nonsensical or actually bad for the health of the country and its citizens. They didn't want the electorate to be easily swayed by sweet sounding promises that are not good for the health of the country in the long run. Since they wanted a level of maturity in the voters that will preclude them from voting for demagogues and panderers, they disliked the idea of giving voting rights to everyone in the country once they reached the age of eighteen or twenty-one.

As we said earlier, they were not rooting for installing a king and a queen. Their recommendation was to limit voting rights to mature knowledgeable individuals only. They felt leaders elected by such voters can and will be able to rule the country well. There is a particular reason that pushed them to think this way. Despite all the name and fame he had, do you know how Socrates died in the end? There was a show trial that examined a bunch of ridiculous charges made against him. There was a jury of one hundred commoners that listened to the charges and arguments, without understanding the details much, and they decided that he should be put to death. He was subsequently executed. So, Plato also was very apprehensive of placing the governance of the whole country in the hands of such an uneducated bunch of idiots.

In the models of democracy we currently practice in the twenty-first century, there is direct democracy and representational democracy. In direct democracy, citizens directly vote on issues to make collective decisions. These are usually referred to as referendums. Whether the UK should leave the European Union or not is one of the recent famous referendums that we all remember. As you may recall, while all the experts and polls predicted that the referendum will show that the UK people want to remain in the EU, a narrow victory went to the Brexit campaign, shocking the world. There were several reports indicating that many who voted to leave didn't really understand what they were voting for or the consequences but just voted to leave thinking they were sticking it to the establishment. Apparently even when they voted to leave, they expected the 'Remain with EU' campaign to win. So, the day the shocking results came out many were Googling such phrases as "What is EU?", "Consequences of Brexit", etc. indicating that these votes may not have been cast after a well thought out deliberation.

In representational democracy, citizens elect their representatives. Once elected, they expect and presume that the representatives will keep the wishes and needs of their constituents and help drive the legislative process to benefit them appropriately. You must be aware of the shock the entire world felt when Donald Trump was declared the winner of the 2016 US Presidential election. His election could be equated to the sweets vendor Plato was talking about in his thought experiment. After winning the presidency telling a large number of voters what they wanted to hear, many fear that he will not prescribe the tough medication the country needs to take it in the right direction. If he continues to pander to his base by going ahead with the sweet delivery, experts wonder what will happen to the patient in the end. The Brexit and the Trump election makes you wonder if the suspicion Plato had about the intelligence of the general public to make tough decisions that will benefit the society and help it move in the right direction is indeed fully justified even today.

Does that mean the current form of democracy we practice where every citizen in the country gets a vote is not a good idea? If we start to look around, we will quickly realize that voting rights have varied in many different ways all over the world all the time. For an example, when the people of India under British rule were given limited voting rights to choose some state level local representatives, not everyone in the country got the right to vote. Only men who owned some land and possessed some level of literacy were initially allowed to vote. As time passed, it was extended to the wives of men who had the right, then to even those without land, subsequently to those who may not be literate, and so on. Only after independence was the Indian constitution updated to provide voting rights to all citizens in 1950. In this century, the voting age was reduced to eighteen allowing a lot more people to vote. I have heard of my father saying elderly Indian women in the 1950's used to say things like they don't have voting rights since they didn't possess land and so had to be educated about the universal franchise promulgated in the independent India.

Today, the notion of democracy is understood all over the world as every citizen of a country gaining the right to vote once they reach certain age, citizens voting to elect their representatives chosen from themselves, and those representatives running the town/state/country for a specific amount of time after which they either step down or run for re-election. Though this universal understanding is widely respected and implemented, if we are faced with the question Plato posed, we can't say for sure that we have resolved the issues he was worrying about. But saying only people with brains can vote won't work since such a rule will immediately run into trouble as to who are the people with brains, who gets to decide who they are and so on. Though the USA proudly proclaims itself as the oldest democracy as often as it can, its history is full of stories where the Caucasian people of power didn't want to give voting rights to African American citizens and so came up with state level laws that said only those blacks that passed a test could get the right to vote, while making sure the test was too hard for them to pass. Using such techniques, racist governments of yesteryear could effectively disenfranchise blacks for decades. It is obvious that as soon as we let some people decide who gets to vote, they will try to leverage that power to benefit themselves or their kind. This will lead to corrupt regimes. So, giving up on the ideas of selecting people that will receive voting rights, let us stick to the idea of every citizen getting their right to vote as soon as they reach a certain age. Let us then think through how the remaining issues could be resolved within this system of voting.

Modern democracies provide one vote per citizen to cast. When the votes are counted, the candidate who gets the most votes wins. This First-Past-The-Post model is very widely used across the globe in a multitude of countries. It is easy to understand and implement. But the big drawback in this system is very well known. Let us say Adam and Bryan are the two contenders in an election. Before the election, predictions based on surveys say that they are going to receive 40% and 60% votes respectively. Then Adam can quickly bring in a new candidate named Charlie who is his friend into the fray. The sole reason Charlie is running in the election is to split Bryan's votes. So, Charlie is chosen to be identical to Bryan wherever it matters, say in race, religion, gender, caste, political positions, etc. With Charlie's entry into the field, Bryan's sixty percent votes get split into roughly 30% each between Bryan and Charlie. Then Adam will win the race with his forty percent, though he is not the majority favorite. We have seen this technique successfully being deployed decade after decade all over the world.

In order to prevent this type of cheating, many countries in the world, including the US and France, force the top two candidates to run in a second run-off election whenever no candidate receives more than fifty percent of the total votes cast. Since only the top two candidates get to participate in the run-off election, Adam will be forced to compete against either Bryan or Charlie in this second-round election. This will ensure that the shenanigans involved in splitting the vote to win an election won't work in the end. Obviously, this is a clean solution to the vote splitting problem and is also fairly easy to understand. But it does end up costing more in terms of time and money since the election must be effectively run twice to elect one candidate. Run your own thought experiments to see how you will solve the problem.

A few countries, Australia being one example, use another clever technique to overcome this issue of vote splitting. In their model, commonly referred to as Alternate Vote, each citizen still gets just one vote. But instead of simply choosing the best candidate on the ballet when they cast their vote, voters rank the candidates as per their preference, giving a rank of 1 for their most preferred candidate and 2 to the next best candidate and so on. This completes the voting process and so it constitutes only one round of voting. But counting the ballots will continue for several rounds. In the first round of counting, your vote is allotted to the candidate you ranked as the best, i.e. number one. Once the votes are all allotted and counted, if any one candidate secures more than fifty percent of the vote, the election process comes to an end with the candidate that received more than fifty percent of the vote being declared the winner.

If no one gets more than fifty percent of the vote in the first round, counting proceeds to the second round. Let's say there are eight contenders in an election. The candidate who received the least number of votes in the first round and ended up in eighth place will be eliminated from the second round. The votes that candidate had in the first round will be redistributed to the remaining seven candidates based on the second choice those ballots indicate. Say for example, the candidate you ranked number 1 in your vote happens to come up short in the first round of counting, unfortunately landing in eighth place. In the second round of counting, your vote will be added to the candidate you ranked as number 2 in your ballot. Once all the votes your first rank candidate received have been redistributed, updated tallies of the remaining seven candidates are compared to see if anyone has gone past the fifty percent mark. If someone has, then that candidate becomes the winner concluding the election process.

If no one manages to get past the fifty percent mark even after the second round, counting proceeds to the third round. Now the candidate in seventh place is eliminated and their votes redistributed again to the remaining six candidates. Let's say the candidate you ranked first is the one leaving in the first round and the one you ranked in the second place is the one leaving after the second round. Then in the third round, your vote will be allocated to the candidate you ranked as number 3. In other words, removed candidate's votes are reallocated to remaining candidates based on whoever happens to be next in the voter's ranking. Now that the number of candidates has shrunk from eight to six, there is a higher probability that one of them might get past the fifty percent mark. If so, they are declared the winner. If not, the process repeats, successively moving the votes to remaining candidates. If the votes were more or less evenly split among all the eight candidates at first, this process may continue for up to seven rounds. At that point, since there will be only two candidates left in the field, the one who gets more vote wins. Though the vote counting process here is a lot more complicated compared to the more common First-Past-The-Post model, once the public is taught to rank the candidates instead of simply selecting just one from the ballot, the election process gets over with just one election, obviating the need for run-off elections, while simultaneously rendering split the vote techniques useless.

There is another model that uses this same technique in a simplified fashion. In that model called Supplementary Vote, the voter picks just one candidate and then optionally indicates a second candidate of choice. No ranking of all the candidates is required. Even marking the second candidate is optional. Just like the ranking model, if the first round of counting yields a decisive winner who had garnered more than fifty percent of the votes, the election ends. If not, only the top two candidates who received maximum votes are retained for the second round, eliminating everyone else. Then all the votes the eliminated candidates received are examined and assigned to one of the two candidates based on the second candidate of choice indicated by the voters. In this case, the second choice might not be one of the top two candidates. So, many votes will be thrown out. But that can be considered equivalent to what happens in a run-off election. Since this effectively boils down to the run-off model, whoever receives more votes after the reallocation wins the election. But this process eliminates the time and energy that needs to be spent on running an actual run-off election. Apparently, the Mayor of London is elected this way. You can compare and contrast these two models to figure out the pros and cons yourself.

There are additional models in practice or being proposed that could be more efficient and foolproof. But the proposals themselves are not thought experiments. So, if the reader is interested, you can search the web to read a lot more. Let us return to another thought experiment now.

Let us set aside the process of casting a vote in elections being held in the country where we live. What if we don't like the structure, rules and regulations that are in effect in the country? What can you do? Of course, you can become a candidate in the next election yourself, win the election and then try to change the system. That could be a crapshoot. You can try to relocate to some other country that suits your principles and temperament. Even that is not that easy to do. Few people who want change but don't have the patience to get to the system of choice by conventional means have proposed an alternative model, more like a thought experiment.

All these laws, rules and regulations tend to prevail on the land masses forming various countries on this planet. If you skip the land and start venturing into the sea, national boundaries start to vanish after about 20 miles. We will then be on international waters where most national laws won't apply. This thought experiment wonders if we can construct large floating squares or rectangles, take them onto the sea and anchor them in the international waters, thus providing alternative city-states for people to live and work. Since these floating blocks won't come under any existing national laws, those who live there can structure their own models of governance, rules of living and try it out. If a set of such blocks like each other, they can get together forming a larger union of blocks. Citizens of such floating blocks or unions can try to lead a happy satisfying life. After a while, be it few months or years, if they run into issues or just get bored of other adjacent blocks, specific blocks can disengage and start living as a separate block or go join some other union of blocks floating elsewhere. This may sound like just living in ships. But unlike ships, the purpose of these blocks is not to constantly travel but just stay in one place to lead a normal life and so costs associated with such a system of floating cities could be much less compared to a traditional ship. Plus, ships are in general designed to be self-contained and not designed to link up with adjacent blocks for a long period of time. The thought here is that such a system will allow experimenting with newer models of governance, education systems, and living style in a more economical way. While this was originally proposed just as a thought experiment, some tech billionaires are apparently trying to fund such projects to see if this can be implemented in reality.

--------- o0o ----------

Teeny Tiny Thoughts

The whole problem with the world is that fools and fanatics are always so certain of themselves, and wiser people so full of doubts.

– Bertrand Russell

Many of the thought experiments we looked at came in with an elaborate background story. But there are also others that are sort of like haikus – they make you think they are just a little riddle or a funny question. As the world delves into them, they turn out to be quite deep leading to interesting results, products and developments in the real world. Wẹ will look at some such experiments in this chapter.

Barber's Paradox

There is a village where the population has one hundred men. Many in that bunch shave their faces themselves. Others reach out to the barber in the village to get shaved. There is only one barber in the village, who is also male. He shaves himself. Now if we separate all the hundred men living in the village into two sets and place all men who shave themselves in Set A and all the men who go to the barber in Set B, does the barber belong to set A or B?

Bertrand Russell

If we change the story a bit where the barber in the village proclaims that he will shave only those men who do not shave themselves, should he shave himself or not? Don't think this is some sort of "Who's on First?" routine done by Abbott & Costello. Circa 1901 Bertrand Russell, who was not only a good philosopher but a well-known philosopher also, came up with this question to point out a flaw in the set theory logic that existed in those days. Mathematicians of those days understood the notion of Set as a list that can be well defined. For a simple example, we can define a set of all even numbers that exist between 1 and 11. Then that set will contain 2, 4, 6, 8, and 10 as its five elements. When Russell thought about this definition of a set, he figured he could define a new more interesting set. His set contained all the sets that are not subsets of themselves. Though it may sound a bit confusing at first, if we look at this definition logically, it is clear and well-defined. So as per the definition of a set at that time, this has to be an acceptable set. A set is always a subset of itself, so this set is part of itself. But as per the definition of this set, it cannot be part of itself. He wondered aloud as to how this paradox can be resolved.

After a lot of people lost a lot of sleep over this question, it forced the mathematicians to redesign the field of set theory itself in mathematics. While many philosophers and mathematicians were trying to answer these questions in a clear binary yes/no format, another school of thinkers thought that instead of trying to find clear answers, we should embrace the innate fluidity these questions possess. This line of thinking leads us to the field of Fuzzy Logic. We should note that the term fuzzy here does not imply confusion but conveys the idea of partial membership to more than one set at the same time. So, the experts that belong to the fuzzy logic group tend not to worry as to whether the barber we met recently belongs to group A or group B but instead give him 0.5 membership to both groups. On a perfunctory glance, this may look like assigning probabilities. But this model of rationally assigning partial membership is fundamentally very different. We are all very familiar with the notion of flipping coins and have heard about the 50% probability assigned to heads & tails each. In that scenario we don't know whether the coin is going to land on its head or tail when flipped. Since it is equally probable that it could be heads or tails, we assign a 50% probability to each possible occurrence. But when we are trying to decide to which group the barber belongs, we have no ambiguity about his membership. We know fully well that he belongs to both groups equally and so give him 0.5 membership to each group. Published literature seems to indicate that Japan has been putting this idea to practical use for a long time now. Starting from designing washing machines to running subway trains, their researchers have been able to successfully use this logic in a wide range of fields and applications.

Fuzzy Logic

"This statement is false". Let's examine this sentence a bit. If that statement is indeed false as it professes, then it becomes a true statement. But if it is a true statement, then as per the statement, it must be a false one. Hmmnn! What is it?

There was a philosopher named Epimenides who lived in Greece circa 500 BC. He made a statement saying, "Cretans are all liars". This created a big paradox since Epimenides himself is from the Greek island of Crete. So, if we consider his statement as true, then since he himself belongs to that island, it will imply that at least one person who is from that island (i.e. Epimenides himself) is not a liar. If so, then that statement is no longer true since he stands as a counterexample to what the statement proclaims. On the other hand, since he is from Crete, if we consider him a liar, then the statement becomes true. Since there is no easy way to get out of this conundrum, it is named after himself as the Epimenides Paradox..

Kurt Gödel

Kurt Gödel is a demigod in the field of theoretical computer science. The Incompleteness Theorem he let loose on this world in the last century is as elegant as Einstein's theory of relativity. The proof he presented flows through mathematics, philosophy and a bunch of other fields like some divine poetry. It is heartening to note that Einstein and Gödel knew each other well and so followed each other's work. If we dig into the proof of the Incompleteness Theorem, we will realize that it starts from an equivalent of a "This statement is false" type point.

There are other equally amusing self-referencing paradoxes floating around like micro thought experiments. Let us look at a couple more. In the Physics chapter, we discussed as to how traveling faster than the speed of light as well as traveling back in time are not possible. But there is a thought experiment that asks what will happen if someone travels back in time and kills one's own grandfather before he had children. If the grandfather didn't have kids, then the grandchild couldn't have been born. This will also imply that the non-existent grandchild couldn't have traveled back in time to kill the grandfather. That means the grandfather would have survived and subsequently could have had children resulting in the birth of this killer grandchild, who could then go back to kill him before he had kids. Though the physical laws of the universe we live in prevents us from traveling backwards and killing our grandparents, this thought experiment has led to several Hollywood movie screen plays, thus putting food on the tables of script writers, actors and directors.

Consider this scenario. A junior attorney joins the well-established law practice of a senior attorney. After learning the ropes of the trade for few years, the junior attorney feels confident enough to hang her own shingle, thus starting a new law practice. As the new practice is being set up, she tells the senior attorney, "As soon as I win my first case, I will pay you the fees for all the things you have taught me over these years". The clever mentor is keen on teaching one last lesson to the mentee and so says she doesn't want to wait till the junior wins a case and wants to be paid right away. To make sure this happens, she promptly sues the junior seeking the fee. If the mentor wins the case, as per the court order the mentee must pay the fee to the mentor. If the mentor loses the case, it will be a victory for the mentee, which means she must now pay the fees to the senior. Either way the process of fee payment gets accelerated as per the senior attorney's wishes. Of course, the mentee, being quite clever herself argues that if she wins, as per the court's decision, she doesn't have to pay the fee. But if she loses the case, she still doesn't have to pay since as per the original contract she hasn't won her first case yet. Known as the Counter dilemma of Euathlus, this paradox is described as a conflict between sophist Protagoras of Greece and his pupil Euathlus.

Two Envelopes

In this experiment, you are a player in a game show. The host has given you two sealed envelopes. One envelope is black in color while the other one is white. Both envelopes contain money, but one has twice the amount inside as the other one.

You just get to pick one envelope and take the money inside. Naturally you want to pick the one with twice the money inside. But since you don't have any idea as to which one has more and which one has less, you are forced to randomly pick one envelope. You can't do any better. So, let us say you pick the white envelope. Before you open the envelope to take the money, the host tells you that, if you like, you can switch and pick the black envelope instead. Should you switch the envelope or stick to the white one you already picked?

White Envelope Black Envelope

You might wonder as to what sort of a stupid question this is! Since you don't know which envelope has how much money, your chance of picking the one with more money is just 50/50. So, all you can do is pick one at random and hope for the best. So what difference is that going to make whether you switch the envelope now or not? Or is it?

Let's analyze this scenario for a minute. Presume the white envelope you chose has $100. Then the black envelope should have either $50 or $200 inside. Since we don't know how much is inside the black envelope, when we chose the black envelope, there is a 50% chance that we will get $50 and there is another 50% chance that we will get $200. So, if you use the probabilities to find what you expect to win by choosing the black envelope, you get ½($200) + ½($50) = $100 + $25 =

$125. This means if you switch the envelope, instead of winning $100 (money inside the white envelope), you are going to win $125 on the average.

Hearing this argument, you feel convinced that it does make sense to switch to the other envelope. So, you switch your selection to the black envelope. Of course, no one has told you how much money is inside even now. So, you can go back to the previous paragraph and think through the same argument. Let's say the black envelope has $100. So, now if you switch to white envelope, it will improve your odds again to win $125 on average. If this logic is correct, it will mean you should keep switching the envelopes forever to improve your odds of landing more prize money. Please try to figure out where the fallacy is in this non-sensical logic. In the next section, we can approach the same problem from a totally different angle.

Monty Hall Problem

Steve Selvin

Professor Steve Selvin of the University of California, Berkeley, posed this brain teaser of a thought experiment circa 1975. Since it resembled the format of a popular "Deal Or No Deal" type game show (called Let's Make a Deal), it came to be known as the Monty Hall problem, based on the TV show host's name.

In this experiment you take part in a TV game show. There are three closed doors in front of you. There is a big prize sitting behind one of the three closed doors. Let's say a million dollars. Other two doors have nothing behind them. Your goal as a contestant is to pick the correct door and score the million-dollar prize. Since all three doors look identical from where you stand, all you can do is hope for the best and pick one door at random. Let's say you choose door one. Once you make your choice, instead of opening door one to show you whether you got the prize or not, the game show host does something weird. He talks about all the stuff you do with a million-dollar prize and then suddenly opens door three, to show you that the prize is not behind door three. So, now you know for sure the prize is behind door one or two.

Now that host offers you a chance to switch your choice. In other words, you can change your selection from door one to door two or continue to stick to door one itself. At this point in the game, will you change your pick to door two, or will you stick to your original selection of door one? Most of the people who hear about this puzzle, don't see any advantage in switching to door two since you still don't know whether the prize is behind door one or two. So, at the most you are running a 50/50 chance of winning the prize. Since you have already selected door one, you still retain the same chance and so audience usually don't see any convincing reason to make the switch that will improve the odds in your favor. Professor Selvin says the right approach is to switch your door selection. But this recommendation appears to be quite counterintuitive. Since we don't know which one of the two doors have the prize, how can simply switching your choice from door one to two improve your odds?

Though the problem was introduced to the public in the 70's, people ranging from Nobel laureates to high school students have continued to fight about the right strategy to improve your odds even in the 1990's. People in the "switch" and "doesn't matter" camps were just about equally divided. Each camp did wonder what's wrong with the other side since their conclusion is so obvious to reach. To finally resolve the issue, game theorists ran simulations on the computer several million times, which conclusively showed that switching improves your odds, forcing the "doesn't matter" camp to switch to the other side eventually, though reluctantly.

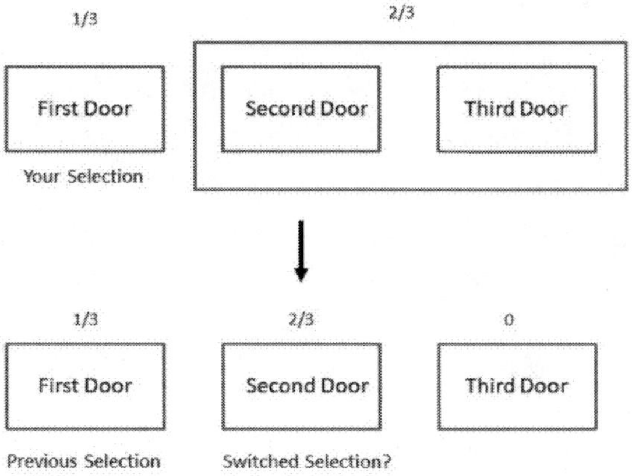

Why the odds would improve if you switch your door selection to the other door, even though both are still closed, can be explained as follows. As shown in the figure below, when you initially choose door 1, it is easy to see that your chance of winning the prize is 1/3rd. Let's say you are allowed to pick doors 2 and 3 together. Then your odds of winning the prize will be 2/3rd. Correct?

At this point, the game show host opens door 3 and shows you that it does not contain the prize. In other words, the odds of winning the prize by choosing door 3 is now proven to be 0. So, the previous 2/3 chance of winning the prize goes behind door 2 alone. The door 1 situation hasn't changed in any way, and so it still retains only 1/3rd odds. So, switching your selection to door 2 will effectively increase your chances to 2/3rd.

Even after this explanation, if you still feel unconvinced, you can think along these lines. Imagine the number of doors used in the show is actually one thousand instead of just three. Initially, you are asked to pick one door out of the one thousand doors. The million-dollar prize is still hidden behind just one door out of those one thousand doors. So, your odds of winning the prize is only 1/1000. Now the game show host steps in and opens not just one door that doesn't have the prize but opens 998 other doors that do not have the prize. So, all you are left with are just two doors, the one you picked and the one the host left unopened. In this scenario won't you switch your selection to the other door? The same analysis and selection strategy will apply to the three doors situation as well.

There is a Hollywood summer Sci-Fi movie called *Interstellar* that came out few years back. In a bleak future portrayed in the movie, earth becomes inhospitable forcing human beings to search for a new home on other planets. After a lot of planning three planets get selected for evaluation as possible locations that are conducive for human beings to relocate. A spaceship is launched to check out the three planets physically. When the explorers traveling in the spaceship analyze and figure out that one of the three planets is not appropriate for relocation, they end up losing a lot more fuel and time than was originally planned. Since they no longer have enough fuel or time to visit the other two planets for evaluation, they are forced to choose only one of the two planets to explore spending their remaining resources. They are thus forced with the challenge of sticking to the same planet they selected originally to visit or switch to the other remaining planet. Naturally in a summer blockbuster movie, this decision will determine the entire fate of the humanity. Though the screen play doesn't call out the dilemma explicitly as the Monty Hall Problem, we can see the similarities in the posed scenario.

--------- o0o --------

Final Round

The human brain has 100 billion neurons, each neuron connected to 10 thousand other neurons. Sitting on your shoulders is the most complicated object in the known universe.

– Michio Kaku

Most of the thought experiments we discussed so far described a scenario and asked us to answer a question or find a path forward. By doing so, they exposed us to different points of view and helped us understand newer concepts.

There are certainly many thought experiments thus structured with a directional motive, trying to nudge the reader into the conclusion the TE designer desires. But at times, some thought experiments simply pose a question and move away without any ulterior motives. The train of thoughts and discussions resulting from such TEs can introduce us to destinations that are not only newer, but we didn't even know existed before. Let us look at one such thought experiment.

Cube-Shaped Earth

Instead of leaving the earth to reach some other destination or look at earth from space as we did in the previous chapters, this thought experiment asks us to imagine our earth itself in a different form. The idea is not new. Many in the past have asked how it would be if Earth is shaped like a cube instead of a sphere. Instead of just wondering vaguely, what will turn up if you approach it as a proper thought experiment and analyze it methodically? Will we be able to walk on the flat surface of the earth till the edge and then jump over to the next face of the cube that is at 90 degrees to the face we have been walking on or will we simply fall off somewhere? Will such a planet earth resemble the current one we occupy in terms of us walking, running, breathing, eating, sleeping on it or will it be a totally different game altogether? It is a very interesting question to delve into properly.

The reason planets tend to be spherical in shape in our universe is the equivalent of the earth's gravitational force that exists on their surfaces due to their mass. If we pick up enough rocks and dirt to put together a cube shaped object that runs several thousand miles in length, width and depth, competing in size with Earth, the gravitational force that forms because of the mass of the object will start from the middle of the object (presuming the density is more or less even across the entire object). This point is the center of gravity. Gravity, as a force will attract inwardly from the object's surface to this point. So, anything and everything on the surface will feel this force and will tend to move towards that center as much as possible. So, over a period of time that could span several million years, that cube shaped collection of objects will morph into a sphere, resembling all the other planets that we see. The notion of Hydrostatic equilibrium will take care of this spherical shape maintenance. If the total size/mass of the object is considerably smaller than the earth, naturally the force of attraction towards its center will be very weak. In that case, if the material with which the object is made is strong enough to resist that weak gravitational force, the total shape can remain in any arbitrary form. This explains why planets of reasonable size tend to be spherical while much smaller comets and other floating objects that are far smaller than planets tend to be in all kinds of random shapes.

Ed Zotti

Since this is only a thought experiment, we have the luxury of imagining whatever we want ignoring the laws of physics. So, let's presume we cemented together a huge cube shaped planet equivalent to the size of Earth. It may tend to become spherical eventually. But we can let our thought process run wild figuring out how the experience of living in such a planet will be thanks to Ed Zotti who has tried to answer this question as scientifically as possible.

Even for a cube-shaped Earth, its center of gravity will be at the center of the cube just like a spherical Earth. When the shape of the earth is a sphere, we can say any point on its surface will be approximately at equidistance from the center of gravity. So wherever we stand on the surface of the Earth, the gravitational force we experience will not change much. But in the case of the cube-shaped earth, how far away a given point on its surface is from its center of gravity will vary dramatically. The eight corners of the cube will be furthest away from the center while the centers of each face (totaling six) will be closest to the center of gravity. This is easy to visualize if you think about the shape of a cube and its center. So, the force of gravity will be at its peak at the centers of each face of the cube and will decrease as you move away from the center of the face towards an edge. It will be least effective at the eight tips of the cube. This variation will lead to several weird consequences.

One of the early and visible impacts will be on how water on the surface of the cube Earth collects. The quantity of available water will get divided into six separate parts with one part occupying each one of the six faces of the cube. So, we will see six separate oceans as shown in the next image. Any fluid that can move easily will encounter the same gravitational pull and so collect the same way on the six faces. We can thus imagine air and steam like substances accumulating like water forming six separate bubbles. In other words the entire atmosphere of each face will be contained in these six bubbles. This will force any lifeforms that depend on oxygen to survive, to live either inside this collection of water and atmosphere bubbles or very close to them. There will be no connection whatsoever between the ocean on one side and the ocean on the other side of the cube. The volume of water found on planet Earth is fairly small compared to its size. Though oceans seem to cover about 71% of Earth's surface, water on the surface is like a thin film compared to its radius. So, the six bubbles that form will be small in size compared to the size of the cube's face itself. If such a planet can exist for a very long time, it would be possible for different life forms to have evolved on each one of those six oceans independently. Since there won't be any contact between those life forms (unless they become technologically advanced enough to travel to the other sides), it would be possible for human beings to live on one side, dinosaurs on the next side, birds on the third, fish & insects on the fourth and so on. Those life forms will continue to thrive without any idea that there are different sets of life forms living on the adjacent sides. This is in a way equivalent to six Earth-like planets getting stitched together to form one large massive cube within our solar system.

Let's say you live in that cube Earth and happen to be an excellent marathon runner. You decide to make a run for one of the eight tips of this cube earth. How will your experience be? Our spherical earth's radius (rounded to the nearest thousand) is about 4000 miles. The length of the equator can be similarly rounded off to 25,000 miles. Since we are imagining a cube shaped earth similar in size, you will probably start from the beach, where the water ends, and dry land begins. This will make you feel like an ant running away from the edge of water stored in a cup and moving towards the cup's rim. Though technically you will be running on a flat surface of the cube, since the gravitational force is pulling you towards the center of the cube at an angle, as you move away from the edge of the ocean towards the edge of the cube, you will feel like you are actually climbing a steep mountain trying to reach its peak.

Let's say in addition to being a major marathon runner, you are also a successful mountaineer who has climbed Mount Everest reaching the summit successfully several times. The biggest mountain peak we know on the surface of the spherical earth, Mount Everest, raises just about 5.5 miles from the sea level. But in a cube-shaped earth, the distance from the center of a face to any one of the four tips of the face will be about 2,800 miles. Comparing the 5.5 mile hike you are used to with the 2,800 mile climb, let's say you decide to skip the idea of running towards and reaching one of the peaks. Instead you decide to take a tour of the equivalent of the equator. In other words, you decide to skip two of the six surfaces of the cube (equivalent to arctic and antartic circles) and walk across the four faces that are adjacent to each other in a long walkathon, similar to traversing the equator. Since you already know that there is no oxygen for you to breath near the corners of the cube, you decide to stay as far away from the tips as possible and walk across the four faces, right in the middle, cutting each face into two halves. This will in effect be walking along the equator of the cube Earth. Since this line is the equivalent of the equator, it will be approximately 25,000 miles long. As you cross each one of the four faces of the cube, you will be forced to cross a large ocean that is sitting in the middle of that face. You get annoyed by the idea of walking through water. So, after some more thought you give up the idea of this long walkathon and decide to hop on a plane to circumnavigate the cube Earth's equator equivalent. Even when you decide to fly, when your plane is trying to fly from one side of the cube to the next side, you will be forced to do maneuvers like flying the plane over a 1582 mile high mountain and then flying down to its base, which will be the middle part of the next side of the cube. The plane will require an enormous amount of fuel to achieve this feat. When we normally fly at 40,000 feet on top of our spherical earth, we are only about 7.5 miles above the earth. It will help to remember that the height of Everest is a mere 5.5 miles.

The size of the water bubbles (i.e. oceans) will be very small compared to the size of the cube's face. So, most of the cube Earth's dry surface will look like unlivable rocky surfaces, devoid of air, that look like mountains sticking into space. In our current spherical Earth, the surface we stand on curves continuously. As you may know, if you stand on the beach and look at the ocean, an ocean liner sailing towards you will show up with its top first at the horizon. As it comes closer to you, you will see more and more of its lower parts. You may recall this effect from geography you studied during school days and perhaps from your actual real-life experience. But since the cube Earth surface not only does not curve but is also devoid of atmosphere in most dry surfaces, we will be able to see things that are very far off without the earth's curvature obstructing our view.

Let's say you ditch even the plane, get on a rocket, launch yourself and reach the line that joins two faces of the cube. Now you look at the two sides of the cube visible to you. Due to lack of gravity that far away from the center of the cube face, you will experience reduced weight and also think you are standing on top of a really tall mountain. You will see the oceans on either side thousands of miles away without any obstruction. As you continue the thought experiment, you will thus land yourself in a strange Earth that is certainly inhospitable to human life.

Beetle in a Box

Ludwig Wittgenstein is a professor of philosophy who lived during the first half of the twentieth century. He was born and brought up in Austria and subsequently served in the University of Cambridge in England. He argued that the meaning of the words in a language as well as the whole language itself are derived from the action of speaking it, without which neither the words nor the language will retain any meaning on their own.

Ludwig Wittgenstein

A bit confusing. Isn't it? Since he himself expected it to be a bit confusing, he proposed a thought experiment to explain the idea. What he is trying to communicate through this TE is how when we try to communicate via language, an experience we encountered through our senses, the thoughts of an individual affects language and understanding.

In this thought experiment, Wittgenstein asks us to imagine a collection of individuals meeting as a group. Everyone in the group has a small box. Inside that box exists something that they describe as a beetle. Each individual is able to see the beetle inside their box but can never see what is inside others' boxes. Now we ask each one to explain the beetle in their box. Irrespective of how well each individual describes his or her beetle, since no one has seen each other's beetles, how much of the explanation reaches the next individual, are there any similarities between the beetle being described and what beetle the listener has, what was exactly understood by the listener are all unknowns. For all we know, one person may have a little bird inside his or her box, the next one may have a piece of chalk and a third one may even have an empty box. If we imagine that group of individuals as a small society, after some time, the term beetle itself might have lost all its ‘a small insect’ type meanings and might be understood as ‘that stuff inside my box’. As an example, the notion of pain is understood by all of us only at an individual experience level. So, we can never be sure if you and I mean the same thing when we talk about pain. So, unless a notion has been learned from a common community experience, what anything truly is becomes unimportant when we describe or debate about it in public. All we can discuss and debate are only those few things that our personal experiences have allowed us to experience and curate in our minds.

If this concept is correct, our private thoughts become the beetle in the thought experiment he proposed. Wittgenstein says that since we can never know or experience the ‘beetle’ of anyone else or, in other words, the private thoughts of another person, there is no guarantee as to whether the language we speak, the thoughts we are trying to express are all reaching the other person as we intended. How each individual views the world, understands what others are saying, interprets the events around them are all dependent on their own personal life experiences and the signals their own senses send into their consciousness. Philosophy professors like to call it Qualia.

Capacity of Thoughts in Human Brain

We are very familiar with medical science replacing and/or restoring some lost functionality of a human body. So, if someone lost a limb, we can provide a prosthesis. If someone’s senses are failing, we can try to restore them. For example, if someone's vision is deteriorating, we can provide glasses or contacts to restore vision. If someone is not able to hear well anymore, we can provide a hearing aid or even a cochlear implant to restore that sense. But are we limited to the five senses we are born with? Is it possible for human beings to acquire new senses that no one else ever possessed? Do a thought experiment. What new sense or power that never existed among human beings would you like?

Dr. Peter König is a German professor/researcher who is interested in answering this question. He picked up a big flat belt and stuck a dozen little motors to it, each looking like a pager. In addition to these motors, this belt also has a GPS device and battery. He asked his test subjects, i.e. his grad students, to wear that belt around their hip for two months. He had rigged it up such that the one motor that is closest to north will keep buzzing like a pager continuously. So, if you wear the belt and standup facing north, the motor near your navel will buzz. If you make a 180 degree turn and face south, the motor on your back will buzz. If you make another 90 degree turn and face west, the motor under your right arm will start buzzing.

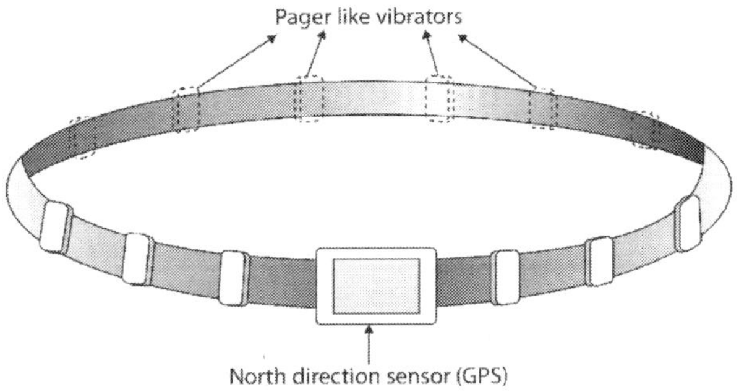

During the first few hours the students wearing this belt felt quite annoyed by the incessant buzzing around their hips. But within a day they got quite used to it. On the following days when they rode their bicycles to the university, as they made each turn on the road, the perpetual buzzing ran around their hips like a little insect, reminding them where north is! As a result, their brains were becoming aware as to where north is throughout the day. It didn't matter whether they needed this information or not, whether they had their eyes open or closed, whether it was day or night, whether they were sitting in a classroom or riding in a bus/train/plane, their brain had this new input flowing in all the time indicating where north was. This was like a new sense specially provided to them similar to what certain migratory birds possess!

Though they got quite used to this sensation within a week and stopped noticing it consciously, they started becoming very aware of how the road curved as they were driving. When they were simply hanging around watching TV at home, a thought popped into their heads saying, "my university lab is exactly in that direction over there". After two weeks, even when they went to a new city, as they were moving around, a map of that new city started developing in their mind. They developed the confidence to claim they could never get lost even in the new city! If any of them were Muslims, I am sure they would have figured out where Mecca is for their five times a day prayer session without having to ask. After collecting their input for two months, Peter took back their belts as his experiment was over. That's it. Most of them felt a profound sense of loss and confusion as if one of their well-developed senses was suddenly taken away! It took them more than a week to recoup normalcy. They did not retain the newfound ability to sense direction.

What we learn from this neat experiment is that if we somehow gather useful information and transmit it to the brain in a consistent fashion, the brain will figure out how to read the incoming data and start putting it to good use very quickly. We could not only use this idea to significantly improve our existing senses (say improving our sense of smell to the level of a dog) but also use it to provide hitherto unknown new senses to human beings. It clearly proves this plausibility. Thus, if we want to sense an oncoming earthquake or be able to sense ultrasonic sound like dolphins or bats, all we need to do is use appropriate sensors and transducers to convert the input into a human perceivable signal and send it over to the brain. The brain will figure out how to read and use it productively!

To Conclude

Neurologists tell us that the prefrontal cortex part of the human brain provides us with an unusual ability that no other life form possesses in the planet. That special ability is our capacity to simulate any given scenario and evaluate what will happen if that scenario manifests itself in reality, pros and cons of reacting to it in various ways and deriving conclusions via this thought process. In order to do this, we don't need any real-life experience of that scenario. It doesn't matter even if that situation is even physically impossible. Solely our imagination is enough to get us going. While other lifeforms can learn from actual experience and remember what happened in the past and put that knowledge to good use in the future, they are not able to simulate situations they have never been in simply by thinking about them. Thought experiments is an embodiment of this unique human ability.

The last three thought experiments we discussed were selected especially for this final chapter. As Wittgenstein says, if any kind of effective communication amongst human beings is innately extremely difficult, despite the existence of such a big barrier, what we have achieved by harnessing the power of the human brain is simply incredible. If we compare Homo sapiens with the age, power, and size/magnitude of the Universe, we look like inconsequential specks. But using the three-pound grey mush inside our heads, we are able to comprehend galaxies and universes spanning millions of light years. Starting from rethinking how our entire planet will be if it is shaped in the form of a cube instead of a sphere, see how far we are able to travel, imagine, guess, analyze and understand. Even in this little book, see how many disciplines, concepts, notions, times and ideas we managed to touch, reach, analyze and experience! All you need for this spectacular journey is just your power of imagination and enthusiasm to experiment with ideas. Let us continue the journey around the universe.

--------- o0o ----------

Acknowledgements

Most of the contents found in this book were originally published as a series of articles in Tamil on the Solvanam.com web magazine. While most of the Tamil language websites tend to trade only on trivialities related to movies & politics, the Solvanam team has created a space to publish thoughtful pieces on science, society and literature. They don't benefit from it financially (no advertisements) or even require users to create a login ID/password to track their interests, which is commendable in this day and age. My kudos and thanks to the Solvanam editorial team.

Our organization within our company develops and delivers extremely complex microchips along with documentation to our customers who use them to design telecommunication equipment. Whenever we interact with our Asian/European customers, one consistent feedback we get is that the documentation we provide should contain more pictures and diagrams compared to text. They insist that walls and walls of text, even if it is exquisitely written, won't help them understand the concepts without accompanying illustrations. As per that lesson learned in my professional career, I have tried to illustrate the writing with as many drawings and pictures as I could. I hope they make understanding the thought experiments we discuss a lot easier. I would be particularly happy if this book helps high school level students & teachers discuss these topics and exchange new ideas. My thanks are due to the drawing master who drew many of the images, my own brother Sarathy who helped coordinate the development of most of the drawings found in the book, institutions and websites that allowed me to use the images and my friend & colleague Mr. Palani Ramanathan who helped with the eBook formatting process. English being my second language, Professor Jerry Trahan of LSU deserves a lot of credit in ensuring all the missing articles are added, grammar errors are corrected, and the flow is improved considerably when I translated the original Tamil book into English. He patiently read through each chapter, asked me challenging questions, provided pointers to enhance the writing and thus helped deliver a much better book than I could have on my own. I continue to be indebted to him for improving my language and flow since my M.S. days when he was correcting my master's thesis on theoretical computer science in the early 90's to this day. A personal family friend Prajakta Paranjpe helped clean up the language further when the book was getting close to completion.

When the articles were getting published originally in Tamil, a famous writer in Tamil briefly mentioned and appreciated me in a column he was writing. So, when the Tamil book was about to come out, I reached out to him seeking a foreword. He basically said no. I didn't feel bad since we didn't know each other personally and so he was just executing what was his prerogative. When he declined my request, it made me stop and think. I then decided to see how well this little piece of work stands on its own merit, without depending on any celebrity recommendation and so gave up the idea of seeking forewords. Some famous person writing a recommendation will help improve the sale. But this approach made me feel better in the long run. These are little experiences, understandings I wouldn't have had, if I had not tried to publish a book. I am grateful for those learnings.

When the articles were being published in Solvanam, several readers read them regularly and provided good feedback and encouragement. My thanks are due to them as well as my own wife and kids who tolerated my attempts to describe & discuss many of these thought experiments to them over long car rides, in restaurants while waiting for food and during walks.

--------- o0o ---------

Index

Made in the USA
Monee, IL
12 November 2026

8d0eb6eb-99d4-490a-bf31-7729c9c6de33R01